Environmental Education and Advocacy

Environmental education has often blurred the distinction between ecological science and environmental advocacy. Growing public awareness of environmental problems and desire for action may be contributing to this blurring. There is a need to clarify the distinction between the role of ecological science and the role of social and political values for the environment within environmental education. This book addresses this need by examining the changing perspectives of ecology in education and the changing perspectives of education in environmental education. Guidelines are provided for assessing the science and education perspectives within environmental education, along with suggested frameworks for development of programs and resources that integrate current science, education, and action. This book will be of interest to environmental educators, ecologists interested in environmental education, and curriculum and resource developers.

EDWARD A. JOHNSON is a Professor in the Department of Biological Sciences, Director of the Kananaskis Field Stations and G8 Legacy Chair in Wildlife Ecology at the University of Calgary. His research interests include plant population and community dynamics, and the coupling of physical processes to ecological processes with respect to natural disturbances. He has carried out fieldwork all over the world and has been a member of the Natural Sciences and Engineering Research Council of Canada's Network of Centres of Excellence in Sustainable Forest Management for the last 12 years.

MICHAEL J. MAPPIN is the Coordinator of Environmental and Ecology Programs for Schools at the University of Calgary Kananaskis Field Stations. In this role, he has 15 years experience in developing curriculum-based environmental and ecology education field studies, programs, and resources. He has served as 2002 Chair of the Education Section and as a member of the Education and Human Resource Committee (1998–2001) and Ecological Visions Committee (2003–2004) for the Ecological Society of America.

Environmental Education and Advocacy

Changing perspectives of ecology and education

Edited by
EDWARD A. JOHNSON
MICHAEL J. MAPPIN
University of Calgary

CAMBRIDGE
UNIVERSITY PRESS

CAMBRIDGE UNIVERSITY PRESS
Cambridge, New York, Melbourne, Madrid, Cape Town, Singapore, São Paulo, Delhi

Cambridge University Press
The Edinburgh Building, Cambridge CB2 8RU, UK

Published in the United States of America by Cambridge University Press, New York

www.cambridge.org
Information on this title: www.cambridge.org/9780521112390

First published 2005
This digitally printed version 2009

A catalogue record for this publication is available from the British Library

ISBN 978-0-521-82410-1 hardback
ISBN 978-0-521-11239-0 paperback

Contents

Contributors

Susan Barker
Warwick Institute of Education, University of Warwick, Coventry CV4 7AL, United Kingdom and Faculty of Education, University of Alberta, Edmonton, Alberta, Canada T6G 2G5 (as of July 1, 2004).

Alan R. Berkowitz
Institute of Ecosystem Studies, Millbrook, New York, United States of America 12545.

Joanna C. Birch
School of Education and Department of Geography, University of Durham, Durham DH1 3LE, United Kingdom.

Rick Bonney
Cornell Lab of Ornithology, Ithaca, New York, United States of America 14850.

Carol A. Brewer
Division of Biological Sciences, The University of Montana, Missoula, Montana, United States of America 59812-4824.

Joseph R. Des Jardins
Department of Philosophy, College of Saint Benedict, Saint John's University, St. Joseph, Minnesota, United States of America 56374.

John F. Disinger
School of Natural Resources, The Ohio State University, Columbus, Ohio, United States of America 43210-1085.

Mary E. Ford
Institute of Ecosystem Studies, Millbrook, New York, United States of America 12545.

Bill Hammond
College of Arts and Sciences, Florida Gulf Coast University, Fort Meyers, Florida, United States of America 33965-6565.

David L. Haury
Mathematics, Science, and Technology Education, College of Education, Ohio State University, Columbus, Ohio, United States of America 43210.

Bob Jickling
Canadian Journal of Environmental Education & Faculty of Education, Lakehead University, Thunder Bay, Ontario, Canada P7B 5E1.

Edward A. Johnson
Department of Biological Sciences & Kananaskis Field Stations, University of Calgary, Calgary, Alberta, Canada T2N 1N4.

Jurek Kolasa
Department of Biology, McMaster University, Hamilton, Ontario, Canada L8S 4K1.

Stein Dankert Kolstø
Department of Applied Education, University of Bergen, Bergen, Norway N-5020.

Marianne E. Krasny
Department of Natural Resources, Cornell University, Ithaca, New York, United States of America 14853-3001.

Michael J. Mappin
Kananaskis Field Stations, University of Calgary, Calgary, Alberta, Canada T2N 1N4.

Milton McClaren
Faculty of Education, Simon Fraser University, Burnaby, British Columbia, Canada V5A 1S6 & Organizational Leadership and

Learning, Royal Roads University, Victoria, British Columbia, Canada V9B 5Y2.

Joy A. Palmer
School of Education and Pro-Vice Chancellor Office, University of Durham, Durham DH1 1TA, United Kingdom.

Steward T. A. Pickett
Institute of Ecosystem Studies, Millbrook, New York, United States of America 12545.

Deborah A. Simmons
Department of Teaching and Learning, College of Education, Northern Illinois University, DeKalb, Illinois, United States of America 60115.

David R. Slingsby
Education, Training and Careers Committee, British Ecological Society, c/o 18 Westfield Grove, Wakefield WF1 3RS, United Kingdom.

Preface

The editors of this book, one an ecologist and one an educator, notice that natural scientists, social scientists, and educators have been diverging in recent decades on their idea of environmental education. Before the 1960s, environmental education was primarily either nature study and outdoor pursuits or natural sciences. Conservation was generally limited to discussions of preserving natural areas and protecting endangered species, particularly birds and mammals. Since the 1960s, environmental education has shifted away from teaching how ecological systems operate to how economic, social, and civil systems are creating environmental problems. This shift in emphasis has resulted in an increase in advocacy in environmental education. In order to consider this changing perspective of ecology and education in environmental education we have divided this book into four parts.

Part I examines some of the fundamental questions about environmental education: How does one distinguish between the different forms and meanings of "ecology" presented in environmental education? What is the role of ecology in environmental education and what are the current ecological concepts that should be included in environmental education? What is the public and academic understanding of the science of ecology?

Part II explores the following questions: What is the public and academic understanding in environmental education? What is the role of government agencies, non-governmental organizations, non-profit groups, industry, and the professional societies in environmental education?

Part III examines the question: How does one distinguish between environmental education and advocacy? Example standards and practical approaches for assessing the science, social, and education dimensions of environmental education resources are provided.

Part IV provides some practical frameworks and approaches for integrating scientific understanding, education, and social action in environmental education.

We would like to thank the following individuals who reviewed chapters: Robert Bohanan, University of Wisconsin-Madison; Glen Chilton, St. Mary's College; Joseph Des Jardins, College of St. Benedict; Bruce Grant, Widener University; Edgar W. Jenkins, University of Leeds; Sharon Kingsland, John Hopkins University; Marylin Lisowski, Eastern Illinois University; Nadine Lymn, The Ecological Society of America; Jurek Kolasa, McMaster University; Richard Kool, Royal Roads University; Libby McCann, University of Wisconsin-Madison; Milton McClaren, Simon Fraser University & Royal Roads University; Rick Mrazek, University of Lethbridge; Lisa H. Newton, Fairfield University; David Ucko, National Science Foundation; Rick Wilke, University of Wisconsin-Stevens Point.

We would also thank Kiyoko Miyanishi, University of Guelph for reading and commenting on the whole book.

Lastly we thank our editors at Cambridge University Press, Alan Crowden and Maria Murphy, the copy editor, Sheila Kane.

MICHAEL J. MAPPIN AND EDWARD A. JOHNSON

1

Changing perspectives of ecology and education in environmental education

INTRODUCTION

> We have specifically directed our attention toward the problem of gaining insight into the educational issue raised by the play of charge and counter-charge, as individuals and groups accuse each other today of using schools in order to indoctrinate young people.
>
> (Nagai 1976: 109)

One of the major challenges facing environmental education today is the growing public attention and concern that education has become blurred with advocacy and that the environmental content in environmental education is no longer based on sound frameworks of natural and social sciences. These concerns are not new but have become more controversial, affecting the future of environmental education within schools.

The concerns over advocacy generally focus on debates about the purpose and teaching approaches found within environmental education curricula, programs, and resources. Debates over purpose are based in part on rapidly changing academic perspectives and public expectations of what constitutes "good" education. Controversy comes about when diverse interest groups involved in environmental education cannot agree on the purpose of education. In the absence of general agreement, the purpose is determined by the dominant interest groups of the time and may not be clearly or explicitly stated.

The concerns with content generally focus on debates about the role of natural or social sciences, humanities, ethics, and religion within environmental education. Given the complex nature of the debates and wide range of perspectives, it is becoming increasingly difficult for educators and the public to differentiate between education and advocacy. This chapter provides an overview of the educational and public debates over advocacy within environmental education and examines how changing perspectives of education and ecology within environmental education contribute to the academic and public concerns of advocacy.

ENVIRONMENTAL EDUCATION AND ADVOCACY

> Taken at face value, the term environmental education must first be concerned with education and second with content about the environment . . . We must, I will argue, be sure that we are educating rather than advocating a particular environmental view. (Jickling 1991: 170–171)

What is the distinction between education and advocacy in environmental education? And why should it matter? By definition, education can be taken to mean the development of the mind's capabilities and character through acquisition of knowledge and abilities to assess and evaluate this knowledge. Advocacy, on the other hand, infers the act of pleading a cause, or encouraging someone to support, speak, or write in favor of a particular behavior or action. Advocacy moves toward indoctrination when *content* is taken to be self-evident or true (tenets, dogma, or doctrines), the *intent* is to have students believe content regardless of evidence to the contrary, and teaching uses *methods* of unquestioned authority or coercion (Nagai 1976; Winch and Gingell 1999). Advocacy and indoctrination also imply a "right way" to act or behave, based on *a priori* values or beliefs, stifling reasoning and understanding. The next two sections consider the educational and public concerns over indoctrination and advocacy within environmental education, focusing on three components of education: the purpose (intent), place (content), and practice (method).

Educational concerns

Concerns over environmental education and advocacy are not new. In 1974, Tanner identified the major issues concerning environmental education as the appropriate *purpose*, *place*, and *practice* for environmental education in schools. Concerns over *purpose* raised questions about the scope and definition of environmental education. The issue of *place* questioned whether environmental education should be an integrated discipline on its own, or inserted and infused throughout all curricula in order to build an environmental ethic, similar to building an ethic of democracy (Tanner 1974). Also, should environmental education be more than science or conservation education and include political, social, philosophical, religious, and moral implications? The issues of *practice* explored the merits and shortcomings of three teaching approaches when dealing with controversial issues, i.e.,"Hands Off," "Soapbox," or "Balanced Exposure" (Tanner 1974: 91). Although tremendous effort and progress have been made in addressing these issues through educational research and practice, the concerns over the purpose, place, and practice of environmental education persist. Over time, the concern over *education* has become as important as the concern over *environment* within environmental education (van Weelie and Wals 2002).

The *purpose* of environmental education has changed from its scientific base in natural science and conservation to a more social and political perspective. Historically this change in purpose is characterized in the education literature as a change from education *about* and *in* the environment to education *for* the environment (Lucas 1980). Education *about* the environment is generally understood to include nature studies, ecology, conservation, and environmental issues. Education *in* the environment is considered more as an approach to education, using the natural and built environments as objects of study. Education *for* the environment implies application or creation of knowledge for social, civil, and political action. The education literature noted that education "for something" could easily lead to advocacy to advance particular behavior, policy, or ideology. For example, Robottom and Hart (1993) point out that the dominant environmental education perspective of the 1980s was to promote environmentally responsible behavior or behavioral change as the primary goal of environmental education (e.g., Hungerford and Volk 1990). Educating for pre-determined behavioral change was deemed antithetical to education. In the 1990s, education *for* the environment (e.g., Fein 1993) was criticized for becoming an instrument for social change rather than an outcome in and of itself (Jickling and Spork 1998). A similar criticism was raised when "education for sustainable development" policy documents were being developed by national governments as an outcome of Agenda 21 (Jickling 1992; McClaren 1993). In this case, it was argued that education was being used to support particular policy outcomes (i.e., sustainable development as interpreted by each government). Education for a particular purpose could no longer be considered educational (McClaren 1993). However, other authors note that the use of "for" simply explains the purpose of that particular education, which varies according to audience and needs, and nothing more (McKeown and Hopkins 2003).

The second major concern focuses on the *place* (where and how) environmental education should be delivered in schools. Historically, environmental education has been considered more an approach rather than a discrete subject. Consequently, there has been greater focus on inserting and infusing environmental education across the curriculum (Simmons 1989) rather than developing an integrated or interdisciplinary subject. Some authors would argue that the infusion approach is the most effective way to proceed, as "all of education is environmental education" (Orr 1994: 12). However, others have argued that infusion has led to a notion that "everything and nothing is environmental education" and this approach contributes to perceptions of advocacy and activism (Knapp 2000).

In most cases, environmental education appears as an add-on or infused into science curricula and is taught predominantly by science teachers using teaching resources that are mostly science based (Simmons 1989). However, environmental education's place in the science curriculum and science teachers' role in environmental education have been questioned by environmental educators throughout the 1980s and 1990s because science was perceived to limit or constrain the scope of environmental education (e.g. Lucas 1980; Robottom 1993a). Reflecting this changing view, the focus on the environment present in environmental education policy statements of the 1970s (e.g., Belgrade Charter and Tblisi Declaration) shifted to a focus on the social or civil in the policies of the 1990s (e.g. Agenda 21) (McKeown and Hopkins 2003). The shift prompted a change in content focus from ecology, resource management, or preservation of natural environments to a social, political, and economic focus. There was also a shift to make the science curricula more socially relevant by linking science content to environmental issues (e.g., Gough 2002; Hart 2002). As part of this process, it has become apparent that there is also an increasing gap between scientific and environmental educators' interpretations of ecology. Interpretations of ecology within curricula and resources have moved from a basis in scientific concepts, theories, and empiricism to value-laden ecopolitical and ecophilosophical concepts, adding to the advocacy debate over content.

The third major concern is the *practice* and role of educators in environmental education. For example, Jickling (1991: 171) described differences between education and training in environmental education. Education was described as "the acquisition of worthwhile knowledge and understanding" and training as the "acquisition of skills and abilities." Jickling argued that environmental education must reach beyond training for action and place more emphasis on worthwhile content, knowledge, and understanding. Otherwise, he noted, being trained how to act without understanding why would lead to naïve problem solving, but not education. Disinger (1999) also cautioned educators that resources developed by government agencies and NGOs tend to focus on policies and mandates of their sponsors rather than on education as the primary outcome. In some cases, these agencies perceive schools as effective locations for advocating their message, policy, or mandate (Disinger 2001), using education as a social instrument in environmental policymaking (van Weelie and Wals 2002). This is reflected in the "plethora of advocacy-oriented documents and curricula which are presented as educational aids" (Hart *et al.* 1999: 116), where education is perceived by agencies outside of schools as a social strategy to achieve a particular policy (e.g., conservation, sustainable development, biodiversity). It is often forgotten or overlooked by educators and teaching resource developers that such education holds *a priori* values and can be perceived as advocacy, especially when the policy outcomes are implied as the only options available (McClaren 1993).

The educators' role in environmental education is also criticized for values they are and are not developing in students (Kelly 2001). Within environmental education literature, educators are advised to be value-free or value-fair in their presentations and development of resources (Disinger 2001). The value-free approach is based on the notion that any personal, religious, or political value position deemed controversial should not be dealt with through schools, but through family, church, or state (Kelly 2001). The intent of the value-fair approach is that all perspectives to a particular issue should be presented to students, so that students can develop their own perspective. It can be argued that this approach promotes relativism, wherein all perspectives can be interpreted as valid and equal. Therefore, other authors note that educators should communicate and share their personal perspectives with students by taking a "committed impartiality" approach to controversial issues (Fein 1993; Kelly 2001). In other words, educators should clearly express their personal views, but still present all perspectives and model how they have assessed and evaluated differing perspectives themselves. As one might expect, any of these approaches, interpreted or handled poorly or with perspectives left out either through error or omission, can lead to major concerns over indoctrination.

Public concerns

In 1997, the Independent Commission on Environmental Education (ICEE) released a white paper entitled *Building Environmental Literacy for the Next Century*, pointing out that: (1) environmental education was needlessly controversial; (2) factual errors were common in teaching resources; (3) costs, benefits, and tradeoffs in environmental problems were not addressed adequately; and (4) many environmental science textbooks had serious flaws, with superficial coverage of topics and mixing of science with advocacy (ICEE 1997; Salmon 2000). The ICEE report focused primarily on content issues and recommended placing greater emphasis on acquisition of natural and social science knowledge, the integration of science, social science, and humanities, and the development of K-12 content standards for environmental studies (ICEE 1997; Disinger 1997). The report was criticized for its overemphasis on developing scientific literacy and lack of emphasis on developing knowledge and skills in civics leading to responsible citizen action, a primary goal of environmental education (Holsman 2001).

In 1996, *Facts, not Fear* by Sanera and Shaw questioned environmental education resources for misuse of facts and for scaring children with misinformation (Sanera and Shaw 1996; Sanera 1998). The book in turn was seriously criticized for its use of flawed research methods (Courtenay-Hall 1998; Simmons 1998) and overemphasis of science (Bowers 1998; Smith 1998) as part of a political and ideological attack on environmental education (Holsman 2001). As *Facts, not Fear* showed, the public debate on environmental education was becoming increasingly politicized. For example, during the reauthorization hearings of the National Environmental Education Act in the United States a representative of the US Environmental Protection Agency, the agency responsible for grant administration, was questioned for using funding under the Act to "brainwash students with environmental teachings" (Education USA 2000). This debate highlighted the differences in public support for environmental education, as was noted years before by Tanner (1974). This intensity of concern with advocacy in schools is likely a reflection of the ebb and flow of liberal and conservative perspectives on public policies and expectations for education at any given time.

The major concern in both education and the public with environmental education has been over educational process and the content of environmental education. In public debates, educators are perceived to lean "away from advocacy" when focusing on knowledge acquisition and processes and to "lean toward advocacy" when focusing on particular values, behaviors, and actions as outcomes (Jickling 2003: 24). Educators themselves hold a wide range of viewpoints on the role of science and advocacy in environmental education as illustrated through opinion editorials in the 1995–97 North American Association for Environmental Education (NAAEE) newsletter, *Environmental Communicator*. Viewpoints range from perspectives that science needs to serve as a foundation to a view that science limits and constrains the intent of environmental education. Opinions also range from advocacy having no place in environmental education to advocacy as a fundamental component of environmental education.

Conflicting perspectives

The controversy within education and the public debates reflect the conflicting perspectives and priorities held by educators, society, and policymakers on what schools should teach, for what ends, and for what reasons (Eisner and Vallance 1974; Beyer and Liston 1996). Eisner and Vallance suggested that the intensity of conflict and difficulty in resolving conflicting perspectives represents a failure of groups to recognize each other's different concept of curriculum and, further, that education is constantly changing, reflecting broader social, political, cultural, and economic priorities of the time (Beyer and Liston 1996). These differing perspectives and changing public priorities inform and determine which purpose, place, and practice are valued at any given time (Eisner and Vallance 1974). It is likely that the conflicting perspectives and priorities arose in part because of a lack of curriculum, standards, scope, and sequence for environmental education until the late 1990s.

During the latter half of the 1990s, the NAAEE developed the *Guidelines for Excellence* for environmental education, as part of the standards movement in education. The guidelines help educators, agencies, and the public develop, assess, and evaluate environmental education programs and resources. However, by necessity, standards need to be framed around a particular educational perspective in order to establish assessment parameters and indicators. The NAAEE guidelines reflect the dominant perspective of environmental educators within the United States of "environmental literacy." Consequently, the guidelines were criticized for constraining and limiting innovation within the field by not incorporating multiple perspectives of environmental education (McClaren 1997; Wals and van der Leij 1997a, 1997b; Hart *et al.* 1999).

Thus, the distinction between education and advocacy in environmental education is much more complex than simply distinguishing between definitions and developing standards. The distinction must place the educational and public debates within the context of changing perspectives and content of environmental education. As noted by

Schubert (1986: 2), when problems are encountered in education, educators need to revisit "perspectives, paradigms and possibilities" of curriculum development in order to become better decision-makers in practice. Schubert noted that perspectives shape our vision of what education ought to be and frame what one accepts or rejects as worthwhile educational content. Perspective can also become ideology when the purpose, content, or approaches to practice are taken for granted as uncontroversial, or even self-evident, and other perspectives are deemed irrelevant or ill informed (Robottom and Hart 1995). For example, phrases or slogans generated from different perspectives of environmental education can be adopted uncritically as the primary purpose of environmental education by agencies or educators to support or amplify their message (Disinger 2001) (e.g. "facts, not fear," "education based in fact, not values," "from knowledge to action," "education for the environment"). The phrase "ecological education" has also taken on multiple and conflicting meanings with different perspectives, sometimes no longer compatible with contemporary concepts of the science of ecology. These slogans take on different meanings from their original sources (Lucas 1995; Jickling and Spork 1998) and become themselves part of the advocacy debate. In short, perspectives can become ideological when there are fixed unquestioned assumptions and *a priori* values of what is right/wrong, good/bad, or moral/amoral, thereby narrowing the conception of what environmental education should or ought to be (Scott and Oulton 1999).

CHANGING PERSPECTIVES OF EDUCATION AND ECOLOGY IN ENVIRONMENTAL EDUCATION

Environmental education has undergone rapid growth and change, not only in terms of purpose, content, and practice, but also in its underlying theoretical foundations or perspectives from research. Some education researchers consider environmental education a young field that still has contested perspectives of what direction it should or will take (Hart *et al.* 1999), while others consider the field to have a well-established framework (e.g. Roth 1997). However, perspectives regarding the purpose for environmental education range from a focus on acquiring disciplinary or interdisciplinary knowledge to changing attitudes, behavior, or values, constructing personal meaning through a reconnection with the natural world, and empowering students to change their world. Environmental education's traditional grounding in nature, science, and conservation has now expanded to include emphases on individual or community-based action, personal spiritual or aesthetic understandings, and critical analysis of values, and natural and social systems (Hart *et al.* 1999).

Environmental education has become education for *behavioral change* (environmentally responsible behavior), *personal change* (enlightenment), and *social change* (emancipation). These perspectives have been developed through different disciplinary lenses or conceptual frameworks that interpret education and the environment in different ways (Smyth 1995). These differences in perspective are reflected in the different priorities and values placed on natural or social sciences and the social objectives for environmental education. Each of the perspectives also draws upon and interprets the science of ecology in different ways in order to help shape its purpose, content, and practice.

Ecology has been part of environmental education throughout its history. Although it is often assumed that ecology has a well-established framework within environmental education (e.g., McKeown and Dendinger 2000), interpretations of ecological concepts can be quite out of date or no longer have the same meaning as in the science of ecology. Ecology has escaped its academic cage (Lambert 1966) and now encompasses multiple meanings and roles outside of science (Westoby 1997). Eco, ecology, and ecological terms have exploded throughout management, education, and popular literature (Wali 1995, 1999) and have taken on intended and unintended value-laden connotations, contributing to the advocacy debate.

Ecology is often equated with environmentalism or "ecological-like thinking" (Kellert and Golley 2000). This may be due to the historical trends of ecology using terms from everyday language to describe what appear to be commonplace phenomena (Lambert 1966). This is apparent in the growing environmental education popular literature, where notions of ecology may not be based in the science of ecology but on interpretations of ecological-like thinking in education (e.g., Hutchinson 1998). Not only are there misconceptions of ecological concepts among students (Munson 1994), but also there appear to be misconceptions and misinterpretations among program developers, authors, and educators as to the meaning of ecology. Many of us are using and teaching ecological concepts that are out of date, simply incorrect science, or not derived from science at all (e.g., ecosystem health). There appears to be an uncritical transition from the facts of ecology to analogies, metaphors, and symbols of ecology in education and nature writing (Phillips 2003) without an explanation of these new meanings. At times, metaphysical or philosophical conceptions of ecology, such as Deep Ecology, are presented as an innovative advance or new paradigm in the scientific understanding of life (e.g., Capra 1996).

As educators, we need to take responsibility for becoming aware of the different perspectives of education and ecology (science) embedded within environmental education policy statements, standards, and resources. In doing so, we not only become more aware of our own perspectives, but also identify and clarify bias, value-laden messages, and inconsistencies between goals, activities, or contexts for learning (van Weelie and Wals 2002). It also helps us in not becoming "pawns in a struggle between contesting messages" (Hart *et al.* 1999: 116). However, it is difficult for educators in practice to work through the contradictory messages (Scott and Oulton 1999) presented by the wide variety of perspectives held for environmental education: e.g. empiricist, positivist, behaviorist, constructivist, interpretivist, critical, eco-feminist, postmodernist, to name but a few (Reid 1996).

Therefore, the next section focuses on three dominant perspectives of environmental education and discusses how each perspective shapes our understanding and interpretation of the purpose and practice of environmental education. The three major perspectives are: environmental education as behavioral change, personal change, and social change. Although presented separately, the three perspectives represent loose conceptual frameworks that overlap or blend in actual theory and practice (McClaren 1997; Roth 1997). Each perspective will be considered for its educational process and ecological content.

Environmental education as behavioral change

Environmental education through environmental citizenship, responsible environmental behavior or environmental literacy has its roots in the 1970s and dominated environmental education research and practice throughout the 1980s and 1990s (Palmer 1998; Rickinson 2001). This perspective is most familiar to educators in the United States through the development and evaluation of frameworks by educational researchers such as Stapp *et al.* (1988), Stapp and Cox (1974), Hungerford *et al.* (1980), Hungerford and Volk (1990), Disinger and Roth (1992), and Roth (1992). These frameworks focus on developing community investigations and citizenship participation by building environmental knowledge, environmental sensitivity and responsible environmental actions (Hungerford and Volk 2003).

Stapp's framework focused on the development of an environmentally educated, concerned and responsible citizenry through a community problem-solving model (Stapp and Cox 1974; Stapp *et al.* 1998) derived from a social sciences action research framework. The Action Research and Community Problem Solving model was applied in international programs such as GREEN (Global Rivers Environmental Education Network) and extended and refined through major environmental education programs throughout Australia and Europe such as the Environmental and School Initiatives (ENSI) Project (Posch 1993; Robottom 1993b). The Hungerford *et al.* framework focused on development of cognitive skills and sociopolitical understandings in order to develop responsible environmental behavior. The model has four goal levels: (1) building ecological foundations - knowledge about ecological concepts; (2) human perceptions and valuing of environment - conceptual awareness of how behavior affects the environment; (3) developing ability to investigate and evaluate problems - knowledge and skills for issue investigation and evaluation; and (4) skills needed to take necessary action (Hungerford and Volk 1990, 2003). The environmental literacy framework focused on building a "capacity to perceive and interpret the relative health of environmental systems and to take appropriate action to maintain, restore, or improve the health of those systems" (Disinger and Roth 1992: 3; Roth 1992). A combination of these three frameworks is reflected in the North American Association of Environmental Education's *National Project for Excellence in Environmental Education: Guidelines for Learning (K-12)* (NAAEE 1999) as four strands: (1) questioning and analysis skills; (2) knowledge of environmental processes and systems; (3) skills for understanding and addressing environmental issues; and (4) personal and civic responsibility.

In educational debates, these perspectives are criticized for not exploring the aesthetic or spiritual dimensions of environmental education (Ashley 2000a) or helping an individual to learn how they think, feel, or draw meaning from the environment. Other educators have criticized this perspective as contradictory to the purpose of environmental education. For example, Robottom and Hart (1995) contend that if behavioral change is made the primary goal, it denies students control over their own learning process and focuses too much on individual behavioral change rather than on addressing the political reform and collective action required to solve environmental issues.

Educational process

Recognizing that the behavioral change has value-laden implications for what constitutes environmentally responsible behavior can lead to an exploration of questions such as: What is environmentally responsible behavior and who decides? (Uzzell 1999). Exploring this question will likely show that environmental values are a complex mix of economic, political, social, personal, and spiritual values that change over time and in different situations (O'Brien and Guerrier 1995). As such, the exploration of diverse personal and social values can be complex and contentious if not handled with care by an educator, or by an educator with little knowledge in the social sciences. In this case, environmental education can be questioned if behavioral change is perceived as the primary goal over knowledge acquisition (Uzzell *et al.* 1995). Other authors, in addition to Uzzell *et al.*, note that this can be alleviated if the focus is on developing decision-making capacity rather than prescribing particular behaviors (Scott and Oulton 1998).

The three frameworks have an action orientation toward management, conservation, preservation, or restoration of the environment as outcomes of responsible environmental behavior. Action is considered a primary component or outcome of these frameworks. However, in actual practice, most educators avoid action projects due to time or curriculum constraints and out of concern for possible community conflict or student failure (Hammond 1997). Environmental education is also perceived to be primarily outdoor or ecological education by many educators and the public. As such, it has been noted that where and when issue investigation and action skills are included, they may not be based on a thorough investigation of alternatives and consequences, resulting in the perception that environmental education is advocacy rather than education (Hungerford 2002). The action component can also be problematic when specific actions are encouraged or presented as ready-made solutions (Uzzell *et al.* 1995). If action and behavioral outcomes are approached or presented as *a priori* values, behavior or action without question, or necessary understanding of the environmental and social consequences of the actions, then it could be argued that this constitutes advocacy rather than education. However, if the recommended actions or behavioral outcomes were explored, assessed, or evaluated in terms of underlying ethics and evidence by the individual or class, then this process "leans away from advocacy toward education" (Jickling 2003).

Ecological content

Both environmentally responsible behavior and environmental literacy regard science and ecology as a foundational component along with civics literacy. Although ecological knowledge is recognized as a prerequisite to sound decision-making, it is regarded as a minor component in shaping environmental citizenship behavior compared to environmental sensitivity, issues knowledge, or action strategy skills (Hungerford and Volk 1990). The ecological concepts focused on are usually population dynamics, nutrient cycling, succession, and homeostasis. However, in practice, programs based on this framework sometimes have difficulty differentiating between natural history, science of ecology, and value-laden management and policy terms (e.g., ecological integrity, ecosystem health).

Many resources use natural history, out-dated conceptions of ecology, or policy-generated terms to serve as the ecological foundation concepts. For example, many field studies are descriptive in nature, making lists of living and non-living things, instead of drawing upon the science of ecology to explain the processes that determine abundance and distribution of those living things. An understanding of the underlying processes is required for management, preservation, conservation, or restoration projects (Pickett *et al.* 1997). Many educational resources still perpetuate out-dated concepts, such as the balance of nature, community or ecosystem stability, or directional notions of succession. Eventually, these can lead to ill-informed public policy and management. Drury (1998) noted that these concepts might have their origin in turn-of-the-century conservation movements that used pre-scientific notions of "Divine Order" or "Nature's Plan" to inform conservation. These notions still persist in conservation and education efforts although they are not based on conservation informed by the contemporary science of ecology (e.g., Pickett *et al.* 1997).

Some educators, non-profits, and management agencies also view conservation biology as the prominent ecological science (e.g. Corcoran 1994) and at times look to that field for guidance and concepts, rather than to ecology, to inform policy and educational resources. Action-oriented and value-laden concepts such as ecosystem health, ecological integrity, and even biodiversity that help set conservation or policy agendas, have particular outcomes in mind (Callicott *et al.* 1999). The terms are intuitively appealing and meet the desire by many for science to provide moral direction and to be action-oriented (i.e. to inform what we should do rather than what is) (Webb 1999). These terms appear frequently in legislation, policies, and educational resources developed by government agencies, NGOs, and educators without precise definition or acknowledgment that they are action-oriented or value-laden. The terms hold *a priori* values, with many having no basis in the science of ecology. A misuse or acceptance without question of these terms only adds to the advocacy debate.

A number of environmental educators are starting to recognize this assumption and are developing programs to help students assess what may be labeled as socioecological terms, such as biodiversity (see Dreyfus *et al.* 1999; van Weelie and Wals 2002). These authors consider concepts such as biodiversity as ill defined in terms of knowledge and value. The terms appear throughout environmental policy, management, education, and popular literature in descriptive (exploring meaning), stipulative (author's definition), or prescriptive (the only meaning) forms. However, terms of this nature do provide a learning opportunity to assess values embedded within resources or programs (Dreyfus *et al.* 1999; van Weelie and Wals 2002). For example, educators and students could examine how emerging or popular scientific concepts can at times be used uncritically to shape mandates, policy, or legislation and how these concepts are not as clearly defined as sometimes presented.

Environmental education as personal change

In spite of increased environmental awareness and concern for global and local environmental issues, there is little evidence of increased environmentally responsible behavior among students or the public (Gigliotti 1990; Uzzell 1999; Ashley 2000a). An alternate approach to increasing environmental responsibility is to focus on developing and nurturing people's understanding of their personal motivations or attitudes that guide their decisions (Neuman 1997). A central premise of this perspective is that social reality exists as people experience it and give meaning to it; reality is individually "constructed, multiple and holistic" (Horton and Hanes 1993: 3). This perspective is referred to as an interpretive approach, education for environmental awareness or interpreting the environment (Fein 1993). The approaches developed from this perspective are generally drawn from the humanities: philosophy, art history, religious studies, linguistics, or literary criticism (Neuman 1997). A few of the current frameworks in this perspective include ecological education (Van Matre and Johnson 1988; 1990; Caduto 1998), Deep Ecology (Devall and Sessions 1985; Naess 1989), and bioregionalism (Smith and Williams 1999).

In ecological education, the major focus is on developing the major ideas of ecology through techniques that make abstract concepts concrete, i.e. an "education of the senses" (Gough 1987; Van Matre and

Johnson 1988; Van Mattre 1990). As interpreted by Caduto (1998), ecological education involves a dynamic, organic, and interactive process that engages multiple intelligences, such as environmental, emotional, intuitive, spiritual, or naturalist intelligence, to connect and establish a healthy relationship with the Earth. The approaches encourage the use of indigenous stories, or traditional wisdom, in addition to the knowledge of science.

Deep Ecology has its roots in the natural philosophy of Arne Naess (1989) and the popular literature of authors such as Devall and Sessions (1985). Deep ecology in the United States is inspired by the nature writings of Aldo Leopold, John Muir, and Henry Thoreau. The intention of this approach is to help individuals develop new perceptions of ecological consciousness, ecological self and life philosophy to help guide everyday living (Gough 1987; Devall 1988). This is based on the premise that "the environmental crisis is a crisis of personal philosophy and requires personal transformation" (Des Jardins 1997). Activities within this approach seek to develop ecological consciousness by reawakening connections with the natural world. As such, the approach relies on not only ecology but also literary works and spiritualism.

Although bioregionalism means different things to different people (Li 2000), most approaches in this framework promote eco-literacy, ecological thinking, and place-based education (Williams 2000). Bioregionalism education seeks to reconnect individuals with their local natural and cultural environments. Additional educational goals include developing a reverence for and understanding of local natural phenomena by becoming aware of ecological relationships and natural laws in one's own region or home-place (Orr 1992; Williams 2000). The relationships and natural laws, in turn, are supported by indigenous or traditional knowledge and cultural norms of the region (Li 2000). This approach also focuses on social change or transformation. Local ecological knowledge is valued and used in education, management, decision-making, and action. Some educational approaches within this framework look to Native American culture and various religions, such as Buddhism, for inspiration and guidance in living and education (Smith and Williams 1996). For example, the Native American belief that one's relationship with creation is connected to one's relationship with the land is seen as helping to build a life philosophy based on natural law and knowledge of place (Scharper 2002).

Educational criticisms of this approach note that there is an over-emphasis on personal transformation rather than social transformation required for sustained social and environmental change (Fein 1993) and that there may be an overemphasis on local concerns at the expense of larger global concerns (Li 2000). These approaches have also been criticized for borrowing practices from various religions out of their cultural or religious context and for being superficial and ineffective in addressing and resolving environmental and social issues (e.g., Bookchin 1991).

Educational process

At times, teaching resources based on these approaches will explicitly, or unknowingly, promote what has been called a "green cosmology" (Scharper 2002), drawing upon philosophy and religion to help guide understanding of the laws of nature, the origins of the world, and our role in it. This goes to the heart of different religious belief structures and can be criticized as indoctrination, animism, or new age religion. Examples of activities that contribute to this criticism include ceremonies such as "Council of all Beings" or uncritical references to "Mother Earth," without thinking what these terms actually mean within a religious context. These approaches may be acceptable to explore within the academic freedom of a university or college setting (e.g., Corcoran 1994) but may not in a school setting.

At another level the frameworks of environmental education as personal change can be prescriptive in providing explicit direction on what we should do or how we should live. For example, some of these approaches take an exclusive eco-centric approach and can lead to concerns of anti-human arguments. Another concern is relativism, i.e. all forms of knowledge and values are held to be valid (e.g., science, religion, personal experience, tradition, common sense, intuition). Consequently, the approach to understanding the world relies heavily on the use of stories, narratives, myths, and rituals (Des Jardins 1997). Relativism and mysticism are not easily accepted by society as a whole, especially within schools.

Ecological content

Environmental education as personal change incorporates metaphysical, spiritual, or religious interpretations of ecology. Thus ecology is used not as science but as a "participatory or sensual experience, relating and connecting to the world from within" (Kirkman 1997). In this second approach, "truths" from ecology are discoverable through "analysis of lived experiences and experiencing of pre-existing truths or natural laws of the world" (Kirkman 1997).

Most of the frameworks are based on an interpretation of nature as complex interwoven relationships, interdependencies, and connectedness (Kirkman 1997). For example, "the health and integrity (or wholeness) of any entity, from cell to organism, from family to community, from classroom to school, from ecosystem to planet, from local economy to global economy, depends on the health of its subsystems and metasystems of which it is a part" (quoted in Sterling 1993: 76). Educators and authors of popular literature in this genre (Sterling 1993; Capra 1996) also note that a new holistic science has emerged that connects psychology, spirituality, theology, and ancient wisdoms with ecological principles of cycles, balance, homeostasis, stability, resilience, complexity, and diversity. They also note that science has failed to keep pace with or use the new advances in holistic science.

These interpretations overgeneralize ecological concepts and ground ethical values in what are perceived as natural laws of ecology (Phillips 2003). The science of ecology does not search for or produce "truth." Therefore, ecological concepts from science should not be used for philosophical or ideological purposes. The transfer of terms from ecology can be misplaced or misleading, but often very intuitive for the public or for educators not versed in science. By moving from ecological fact to analogy and metaphor, ecology becomes a "point of view" rather than a science (Phillips 2003). Ecology can then be taken to be the source of holistic approaches while at the same time scorned as being overly reductionist.

Environmental education as social change

The third major perspective is based on critical social theory that argues that environmental problems are rooted in social, economic, and political systems and the worldviews that support those systems (Fein 1993). This perspective sees environmental education as more than ecology, behavioral change, or interpretation of nature, but as an approach to changing social values and systems in order to achieve sustainability and social justice. The previous two perspectives are seen as overemphasizing a focus on individual behavior and science, rather than on social change (Neuman 1997). This approach emerged in environmental education during the 1990s, especially in education research and practice in Australia (Robottom 1987, 1993b; Fein 1993) and later in Europe

(Posch 1993). Community action groups, political organizations, social movements, and environmental justice groups usually adopt this perspective. The conception that environmental education is unavoidably political also emerges in the popular literature, wherein educators are encouraged to move beyond teaching students existing patterns to a vision of reshaping society toward sustainability (Orr 1992).

Educators from this perspective criticize the environmental education citizenship perspective for failing to address the needs of real people and their social conditions. On the other hand, the interpretive perspective is considered too subjective, and relativistic, and as failing to take a strong value position or to help people change themselves and society (Fein 1993). However, some critical theorists take a more expansive view that all three perspectives are required for ecologically sustainable development (Huckle 1993). The behavioral change and interpretive perspectives seek to describe the world as it is and the critical theory perspective tries to understand why the social world is the way it is and strives to know how it should be (Sterling 1993). In this perspective, the environmental crisis is a social crisis and requires both social transformation and social activism for the environment (Fein 1993).

Therefore, the primary outcomes for this perspective of environmental education are inquiry, emancipation, empowerment, and environmentalism (Huckle 1993). From this perspective, the purpose of environmental education is defined as follows:

> As a professional practice, environmental education seeks to develop the understandings, values and action skills necessary for people to work with others to improve the quality and sustainability of their natural and social environments. Environmental education seeks to provide life-long learning experiences through which people may take a place in society as informed, committed and active citizens who are capable of playing a part in making their society a better place in which to live by caring about the needs of all species, and by speaking out and acting against social and ecological injustice. (Fein 1993: viii)

The criticism of this approach is its value-laden assumptions (Scott and Oulton 1999). Education holding this perspective makes assumptions about the relationship between environmental education and environmental ideologies (green politics, red-green politics), that scientific approaches are behaviorist or positivistic, and considers the social critical approach as the only valid approach for environmental education (Connell 1997; Walker 1997).

Educational process

In this perspective education itself is perceived to be part of the environmental crisis. This social activist orientation considers education a moral-political activity requiring the educator to commit to a value position (Neuman 1997). Few educators in practice understand this perspective for social transformation or reconstruction to be a prime purpose of school. This approach may not be practical for most schools in that schools cannot accommodate the necessary social change required by critical theory (Walker 1997), as some critical theorists go so far as to prescribe preferred ecosocialist political orientation or vision of social change (e.g., Fein 1993; Sterling 1993).

Ecological content

This perspective uses the science of ecology to inform itself of environmental crises, but at times rejects science as contributing to environmental degradation (see e.g., Robottom 1993a). Science and other forms of environmental education are also presented as being reductionist, behaviorist, or positivist in approach, perpetuating "man apart from nature" thinking (Connell 1997). As such, Deep Ecology, ecophilosophy, or metaphysical ecology is usually presented as preferred sources of ecological information. This perspective holds a social interpretation of ecology as holistic, multidisciplinary, and a source for a new world ethic of living. However, others note that this reductionist–holistic dualism has no basis in current scientific thinking (Haila 2000) and therefore should have no basis in education.

BEYOND PERSPECTIVES AND POLITICS

The previous section illustrates how different perspectives of environmental education hold particular philosophical and social purposes for education that, in turn, affect content, interpretations of ecology, and concerns for advocacy within environmental education. As educators we need to move beyond the politics of advocacy in environmental education and address what is philosophically, ethically, and practically valid within schools. We need to ask three fundamental questions: (1) How do we decide which purposes are appropriate or not for environmental education? (Connell 1997). (2) How do we know whether a particular approach will assist individuals in making informed decisions? (Connell 1997). (3) What is the proper goal of science in environmental education (is it explanation, enlightenment, or emancipation)? (Koertge 2000).

In response to the first question, an understanding of the changing perspectives helps us to challenge and debate *a priori* values or ideology present in policy, curricula, resources, or practice. This means that in formal or non-formal education there is a responsibility to become familiar with the value and limitations of the various perspectives of environmental education. There is a place for all three of the major perspectives at particular times and in particular courses. It is our job as educators to ensure that perspectives match our educational goals for a discipline and do not become ideologies. Perspectives need to be innovative, but supportive and relevant to practicing educators, practical to implement, and credible to the public (Smyth 1995).

In order to deal with the second question of assisting students to make informed decisions, we need to get on with the difficult work of developing transdisciplinary and interdisciplinary programs, programs that build fundamental knowledge within disciplines and at the same time provide an opportunity to synthesize that knowledge in response to a local environmental problem that is appropriate for students, a school, and community (Lucas 1985). In practice, there is significant resistance to interdisciplinary approaches across natural and social sciences in school organization (Disinger 1990). However, one approach may be to use environmental education as capstone courses that integrate natural sciences, social sciences, and humanities (Lucas 1985; Disinger 1997, 2001). Every discipline has the potential to contribute to environmental education (Smyth 1995; McKeown and Hopkins 2003). Specific courses would also provide a greater opportunity to link practice with research (Disinger 2001) and ground research in practice (Palmer 2000). We also need to work within existing structures rather than trying to radically change the social structure of schools; otherwise, there is danger of environmental education becoming even more marginalized.

In regard to the third question, ecology as a science is perceived as an "ally and as a foe to environmental education" (Ashley 2000b), even though all the major perspectives use concepts from ecology. There appears to be an increasing gap between ecologists and environmental educators in their interpretations of ecology. Interpretations of ecology used in educational theory and practice vary widely within and between each of the educational perspectives. There are various interpretations of ecology based on natural history, and ecosystem, population, and evolutionary ecology. In addition, a wide array of literary, philosophical, and social interpretations of ecology (e.g., the Land Ethic, Deep Ecology, social ecology, or ecofeminism) serve as the basis for ecology in each educational perspective. The use of the science of ecology as a basis for ethical or philosophical arguments may not be considered valid in philosophy. For example, Des Jardins (1997) notes that similar ecological concepts can lead to different ethical conclusions, and similar ethical conclusions can be drawn from different ecological concepts.

Ecology within the science and environmental education resources and curricula is being replaced with value-laden concepts from management, policy, literature, or social sciences. Koertge (2000) notes that science education is being affected by "new age" ideas, and by groups that have no background in either the philosophy of science or science. This slow change of understanding of ecology, both in terms of content and approach, toward value-laden conceptions will only exasperate the advocacy debate. Conceptions based on interpretations of ecology will only lead to unsound education and decision-making in the future. Educators need to explain the difference between the various interpretations of ecology (Westoby 1997), stipulate meanings and concepts more rigorously, and to develop environmental education frameworks that integrate ecological information and research for problem-solving at local levels (e.g., Castillo *et al.* 2002).

CONCLUSION

> In our society (that is advanced western society) we have lost even the pretense of a common culture. Persons educated with the greatest intensity we know can no longer communicate with each other on the plane of their major intellectual concern. This is serious for our creative, intellectual and, above all, our normal life. It is leading us to interpret the past wrongly, to misjudge the present, and to deny the hopes of the future. It is making it difficult or impossible for us to take good action. (Snow 1964: 60)

This overview of the different perspectives of education and ecology within environmental education illustrates that there is still a gap between the natural sciences, social sciences, and humanities. The gap is now expanding rather than being narrowed by education. For example, Smyth(1995: 8) notes that sometimes those who focus on understanding the environment have difficulty dealing with the human element and those who focus on people know little of the ecological constraints. Smyth (1995) also noted that rapidly changing perspectives of education and ecology only add to this gap and in times of rapid change there is greater susceptibility to advocacy or indoctrination.

It is time to bring together the humanities, natural sciences, and social sciences (Lucas 1985, Smyth 1995; Wilson 1998; Disinger 2001), and to work together through education to develop more accurate interpretations of complex concepts and issues. As noted earlier, simply borrowing concepts from a discipline such as ecology does not portray or do justice to the complexity of the environment, nor to human thought (Kirkman 1997). Nowhere is this more apparent than in current science and environmental education literature.

As educators, we need to be able to distinguish in environmental education between scientific, emotional, ethical, and political arguments (Jickling 1991; van Weelie and Wals 2002). We need to move beyond the debates on perspectives and return to the fundamental question of what it means to be educated:

> To be educated implies certain habits of mind: the formation of rational arguments, appreciation of how various disciplines construct knowledge and assess truth and value, an appreciation of logic, a desire to form one's own views, and a willingness to reflect on those views in the context of differing ideas. To be educated also implies a resistance to propaganda, indoctrination and ideology. (McClaren 1993: 17)

REFERENCES

Ashley, M. (2000a). Behaviour change and environmental citizenship: a case for spiritual development? *International Journal of Children's Spirituality,* **5**(2), 131–145.

(2000b). Science: an unreliable friend to environmental education? *Environmental Education Research,* **6**(3), 269–280.

Beyer, L. E. and Liston, D. P. (1996). *Curriculum in Conflict: Social Visions, Educational Agendas, and Progressive School Reform.* New York, NY: Teachers College Press.

Bookchin, M. (1991). Will ecology become "the dismal science"? *Progressive,* **55**(12), 18–21.

Bowers, C. A. (1998). A cultural approach to environmental education: putting Michael Sanera's ideology into perspective. *Canadian Journal of Environmental Education,* **3**, 57–66.

Caduto, M. J. (1998). Viewpoint. Ecological education: a system rooted in diversity. *Journal of Environmental Education,* **29**(4), 11–16.

Callicott, J. B., Crowder, L. B., and Mumford, K. (1999). Current normative concepts in conservation. *Conservation Biology,* **13**(1), 22–35.

Capra, F. (1996). *The Web of Life: A New Scientific Understanding of Living Systems.* New York, NY: Anchor Books.

Castillo, A., Garcia-Ruvalcaba, S., and Martinez, L. M. (2002). Environmental education as a facilitator of the use of ecological information: a case study in Mexico. *Environmental Education Research*, **8**(4), 395–411.

Connell, S. (1997). Empirical-analytical methological research in environmental education: response to a negative trend in methodological and ideological discussions. *Environmental Education Research*, **3**(2), 117–132.

Corcoran, P. B. (1994). Reconceptualizing environmental education: five possibilities. *Journal of Environmental Education*, **25**(4), 4–8.

Courtenay-Hall, P. (1998). Textbooks, teachers and full-colour vision: some thoughts on evaluating environmental education "performance." *Canadian Journal of Environmental Education*, **3**, 27–40.

Des Jardins, J. R. (1997). *Environmental Ethics: An Introduction to Environmental Philosophy*, 2nd edn. Belmont, CA: Wadsworth Publishing Company.

Devall, B. (1988). *Simple in Means, Rich in Ends: Practicing Deep Ecology*. Salt Lake City, UT: Gibbs Smith.

Devall, B. and Sessions, G. (1985). *Deep Ecology: Living as if Nature Matters*. Salt Lake City, UT: Peregrine Smith.

Disinger, J. F. (1990). Environmental education for sustainable development? *Journal of Environmental Education*, **21**(4), 3–6.

(1997). Are we building environmental literacy? A report on this report. *Environmental Communicator*, **27**(3), 15.

(1999). Environment in the K-12 curriculum: an overview. In *Environmental Education Teacher Resource Handbook: A Practical Guide for K-12 Environmental Education*, ed. R. J. Wilke. Kraus International Publications, pp. 23–43.

(2001). K-12 education and the environment: perspectives, expectations, and practice. *Journal of Environmental Education*, **33**(1), 4–11.

Disinger, J. F. and Roth, C. E. (1992). Environmental Literacy. ERIC: CSMEE Digest. ED351201.html

Dreyfus, A., Wals, A. E. J., and van Weelie, D. (1999). Biodiversity as a postmodern theme for environmental education. *Canadian Journal of Environmental Education*, **4**, 155–175.

Drury Jr., W. H. (1998). *Chance and Change: Ecology for Conservationists*, ed. J. G. T. Anderson. Berkley, CA: University of California Press.

Education USA. (2000). EPA's environmental teaching causes partisan feud. Education USA, pg. 12, July 10.

Eisner, E. W. and Vallance, E., eds. (1974). *Conflicting Conceptions of Curriculum*. Berkeley, CA: McCutchan Publishing Corporation.

Fein, J., ed. (1993). *Environmental Education: A Pathway to Sustainability*. Geelong, Australia: Deakin University Press.

Gigliotti, L. M. (1990). Viewpoint. Environmental education: what went wrong? What can be done? *Journal of Environmental Education*, **22**(1), 9–12.

Gough, A. (2002). Mutualism: a different agenda for environmental and science education. *International Journal of Science Education*, **24**(11), 1201–1215.

Gough, N. (1987). Learning with environments: towards an ecological paradigm for education. In *Environmental Education: Practice and Possibility*, ed. I. Robbottom. Victoria, Australia: Deakin University Press, pp. 49–67.

Haila, Y. (2000). Beyond the nature-culture dualism. *Biology and Philosophy*, **15**, 155–175.

Hammond, W. F. (1997). Educating for action: a framework for thinking about the place of action in environmental education. *Green Teacher*, **50**, 6–14.

Hart, P. (2002). Environment in the science curriculum: the politics of change in the Pan-Canadian science curriculum development process. *International Journal of Science Education*, **24**(1), 1239–1254.

Hart, P., Jickling, B., and Kool, R. (1999). Starting points: questions of quality in environmental education. *Canadian Journal of Environmental Education*, **4**, 104–124.

Holsman, R. H. (2001). Viewpoint. The politics of environmental education. *Journal of Environmental Education*, **32**(2), 4–7.

Horton, R. L. and Hanes, S. (1993). Philosophical considerations for curriculum development in environmental education. ERIC: CSMEE Bulletin SEB93-4.

Huckle, J. (1993). Environmental education and sustainability: a view from critical theory. In *Environmental Education: A Pathway to Sustainability*, ed. J. Fein. Geelong, Australia: Deakin University Press, pp. 43–68.

Hungerford, H. R. (2002). Environmental educators: a conversation with Rick Wilke. *Journal of Environmental Education*, **33**(4), 4–9.

Hungerford, H. R., Peyton, R. B., and Wilke, R. J. (1980). Goals for curriculum development in environmental education. *Journal of Environmental Education*, **11**(3), 42–47.

Hungerford, H. R. and Volk, T. L. (1990). Changing learner behavior through environmental education. *Journal of Environmental Education*, **21**(3), 8–21.

Hungerford, H. R. and Volk, T. L. (2003). Notes from Harold Hungerford and Trudi Volk: meeting the challenges facing the environmental education community. *Journal of Environmental Education* **34**(2), 4–6.

Hutchinson, D. (1998). *Growing up Green: Education for Ecological Renewal*. New York, NY: Teachers College Press.

Independent Commission on Environmental Education. (1997). *Are We Building Environmental Literacy?* Washington, DC: ICEE.

Jickling, B. (1991). Environmental education and environmental advocacy: the need for a proper distinction. In *To See Ourselves/To Save Ourselves: Ecology and Culture in Canada*, ed. R. Lormimer, M. McGonigle, J. -P. Reveret, and S. Ross. Montreal: Association for Canadian Studies, pp. 169–176.

(1992). Why I don't want my children to be educated for sustainable development. *Journal of Environmental Education*, **23**(4), 5–8.

(2003). Environmental education and environmental advocacy: revisited. *Journal of Environmental Education*, **34**(2), 20–27.

Jickling, B. and Spork, H. (1998). Education for the environment: a critique. *Environmental Education Research*, **4**(3), 309–327.

Keller, D. R. and Golley, F. B., ed. (2000). *The Philosophy of Ecology: From Science to Synthesis*. Athens, GA: The University of Georgia Press.

Kelly, T. E. (2001). Discussing controversial issues: Four perspectives on the teacher's role. In *Philosophy of Education: Introductory Reading*, ed., W. Hare and J. P. Portelli, 3rd edn. Calgary, Canada: Detselig Enterprises Ltd., pp. 221–242.

Kirkman, R. (1997). Why ecology cannot be all things to all people: the "adaptive radiation" of scientific concepts. *Environmental Ethics*, **18**, 375–390.

Knapp, D. (2000). The Thessaloniki Declaration: a wake-up call for environmental education? *Journal of Environmental Education*, **31**(3), 32–39.

Koertge, N. (2000). "New age" philosophies of science: constructivism, feminism and postmodernism. *British Journal for the Philosophy of Science*, **51**(4), 667–683.

Lambert, J. M., ed. (1966). *The Teaching of Ecology: A Symposium of The British Ecological Society*, April 13-16. University of London. Oxford: Blackwell Scientific Publications.

Li, H. (2000). Bioregionalism and global education: exploring the connections. *Philosophy of Education*, 394–401.

Lucas, A. M. (1980). Science and environmental education: pious hopes, self praise and disciplinary chauvinism. *Studies in Science Education*, **7**, 1–26.

(1985). Environmental education: what is it, for whom, for what purpose, and how? In *Conceptual Issues in Environmental Education*, ed. S. Keiny and U. Zoller. New York, NY: Peter Lang, pp. 25–48.

(1995). *Beware of slogans! Environmental Education: From Policy to Practice*. British Council Seminar, King's College London.

McClaren, M. (1993). Education, not ideology. *Green Teacher*, **35**, 17–18.

(1997). Reflections on "Alternatives to National Standards in Environmental Education: Process-Based Quality Assessment." *Canadian Journal of Environmental Education*, **2**, 35–46.

McKeown, R. and Dendinger, R. (2000). Socio-political-cultural foundations of environmental education. *Journal of Environmental Education*, **31**(4), 37–45.

McKeown, R. and Hopkins, C. (2003). EE≠ESD: Defusing the worry. *Environmental Education Research*, **9**(1), 117–128.

Munson, B. H. (1994). Ecological misconceptions. *Journal of Environmental Education*, **25**(4), 30–34.

Naess, A. (1989). *Ecology, Community and Lifestyle*, ed. D. Rothenberg. Cambridge: Cambridge University Press.

Nagai, M. (1976). *Education and Indoctrination: The Sociological and Philosophical Bases*. Tokyo, Japan: University of Tokyo Press.

Neuman, W. L. (1997). *Social Research Methods: Qualitative and Quantitative Approaches*, 3rd edn. Needham Heights, MA: Allyn & Bacon.

North American Association for Environmental Education. 1999 (2000). *Excellence in Environmental Education: Guidelines for Learning (K-12)*. Rock Spring, GA: NAAEE.

O'Brien, M. and Guerrier, Y. (1995). Values and the environment: an introduction. In *Values and the Environment: A Social Science Perspective*, ed. Y. Guerrier, N. Alexander, J. Chase, and M. O'Brien. Chichester: John Wiley and Sons, pp. xiii–xvii.

Orr, D. W. (1992). *Ecological Literacy: Education and the Transition to a Postmodern World*. Albany, NY: State University of New York Press.

(1994). *Earth in Mind: On Education, Environment, and the Human Prospect*. Washington, DC: Island Press.

Palmer, J. A. (1998). *Environmental Education in the 21st Century: Theory, Practice, Progress and Promise*. London: Routledge.

(2000). Research matters: a call for the application of empirical evidence to the task of improving the quality and impact of environmental education. *Cambridge Journal of Education*, **29**(3): 379–395.

Phillips, D. (2003). *The Truth of Ecology: Nature, Culture, and Literature in America*. New York, NY: Oxford University Press.

Pickett, S. T. A., Ostfeld, R. S., Shachak, M., and Likens, G. E., eds. (1997). *The Ecological Basis of Conservation: Heterogeneity, Ecosystems and Biodiversity*. New York, NY: Chapman & Hall.

Posch, P. (1993). Research issues in environmental education. *Studies in Science Education*, **21**, 21–48.

Reid, A. (1996). Signposts in a strange land: a response to Peter Posch. *Environmental Education Research*, **2**(3), 363–372.

Rickinson, M. (2001). Learners and learning in environmental education: a critical review of the evidence. *Environmental Education Research*, **7**(3): 207–320.

Robottom, I., ed. (1987). *Environmental Education: Practice and Possibilities.* Victoria, Australia: Deakin University.

Robottom, I. (1993a). The role of ecology in education: an Australian perspective. In *Ecology in Education,* ed. M. Hale. Cambridge: Cambridge University Press, pp. 1–9.

(1993b). *Policy, Practice, Professional Development and Participatory Research: Supporting Environmental Initiatives in Australian Schools.* Victoria, Australia: Deakin University.

Robottom, I. and Hart, P. (1993). *Research in Environmental Education: Engaging the Debate.* Victoria, Australia: Deakin University.

(1995). Behaviorist EE research: environmentalism as individualism. *The Journal of Environmental Education,* **26**(2), 5–9.

Roth, C. E. (1992). Environmental literacy: its roots, evolution and directions in the 1990s. ERIC: CSMEE ED348235.

Roth, R. E. (1997). A critique of "Alternatives to National Standards for Environmental Education: Process-Based Quality Assessment." *Canadian Journal of Environ-mental Education,* **2**, 28–34.

Salmon, J. (2000). Are we building environmental literacy? *Journal of Environmental Education,* **31**(4), 4–10.

Sanera, M. (1998). Environmental education: promise and performance. *Canadian Journal of Environmental Education,* **3**, 9–26.

Sanera, M. and Shaw, J. S. (1996). *Facts, not Fear: A Parent's Guide to Teaching Children about the Environment.* Washington, DC: Regnery.

Scharper, S. B. (2002). Green dreams: religious cosmologies and environmental commitments. *Bulletin of Science, Technology & Society,* **22**(1), 42–44.

Schubert, W. H. (1986). *Curriculum: Perspective, Paradigm, and Possibility.* New York, NY: Macmillan Publishing Company.

Scott, W. and Oulton, C. (1998). Environmental values education: an exploration of its role in the school curriculum. *Journal of Moral Education,* **27**(2), 209–224.

(1999). Environmental education: arguing the case for multiple approaches. *Educational Studies,* **25**(1): 89–97.

Simmons, D. A. (1989). More infusion confusion: a look at environmental education curriculum materials. *Journal of Environmental Education,* **20**(4), 15–18.

(1998). Reflections on "Environmental Education: Promise and Performance." *Canadian Journal of Environmental Education,* **3**, 41–47.

Smith, G. (1998). Response to "Environmental Education: Promise and Performance." *Canadian Journal of Environmental Education,* **3**, 48–55.

Smith, G. and Williams, D. R. (1996). The greening of pedagogy: reflections on balancing hope and despair. *Holistic Education Review,* **9**(1), 43–51.

Smith, G. and Williams, D. R., eds. (1999). *Ecological Education in Action: On Weaving Education, Culture, and the Environment.* Albany, NY: State University of New York.

Smyth, J. C. (1995). Environment and education: a view of a changing scene. *Environmental Education Research,* **1**(1), 3–20.

Snow, C. P. (1964). *The Two Cultures: And a Second Look.* Cambridge: Cambridge University Press.

Stapp, W. and Cox, D. A., ed. (1974). *Environmental Education Activities Manual. Book 1: Concerning Spaceship Earth.* Farmington Hills, MI: Stapp and Cox.

Stapp, W., Bull, J., Cromwell, M., Cwikiel, J. W., Di Chiro, G., Guarino, J., Rathje, R., Wals, A., and Youngquist, M. (1988). *Education in Action: A Community Problem Solving Program for Schools.* Dexter, MI: Thomson-Shore Inc.

Sterling, S. (1993). Environmental education and sustainability: a view from holistic ethics. In *Environmental Education: A Pathway to Sustainability*, ed. J. Fein. Geelong, Australia: Deakin University Press, pp. 69-98.

Tanner, R. T. (1974). *Ecology, Environment, and Education.* Lincoln, NE: Professional Educators Publications.

Uzzell, D. (1999). Education for environmental action in the community: new roles and relationships. *Cambridge Journal of Education*, **29**(3), 397–413.

Uzzell, D., Rutland, A., and Whistance, D. (1995). Questioning values in environmental education. In *Values and the Environment: A Social Science Perspective*, ed. Y. Guerrier, N. Alexander, J. Chase and M. O'Brien. Chichester: John Wiley, pp. 171–182.

Van Matre, S. (1990). *Earth Education ... A New Beginning.* Warrenville, IL: The Institute for Earth Education.

Van Matre, S. and Johnson, B. (1988). *Earthkeepers$_{TM}$: Four Keys for Helping Young People Live in Harmony with the Earth.* Warrenville, IL: The Institute for Earth Education.

van Weelie, D. and Wals, A. E. J. (2002). Making biodiversity meaningful through environmental education. *International Journal of Science Education*, **24**(11), 1143–1156.

Wali, M. K. (1995). Ecovocabulary: a glossary of our times. *Bulletin of the Ecological Society of America*, **75**, 106–111.

(1999). Ecology today: beyond the bounds of science. *Nature and Resources*, **35**, 38–50.

Walker, K. (1997). Challenging critical theory in environmental education. *Environmental Education Research*, **3**(2), 155–162.

Wals, A. E. J. and van der Leij, T. (1997a). Alternatives to the National Standards for Environmental Education: process-based quality assessment. *Canadian Journal of Environmental Education*, **2**, 7–27.

(1997b). Alternatives to the National Standards for Environmental Education: a response to Roth and McClaren. *Canadian Journal of Environmental Education*, **2**, 49–57.

Webb, N. R. (1999). Ecology and ethics. *Trends in Ecology and Evolution*, **14**(7), 259–260.

Westoby, M. (1997). What does "ecology" mean? *Trends in Ecology and Evolution*, **12**(4), 166.

Williams, D. R. (2000). Re-coupling place and time: bioregionalism's hope for situated education. *Philosophy of Education*, 404–407.

Wilson, E. O. (1998). *Consilience: The Unity of Knowledge.* New York, NY: Vintage Books.

Winch, C. and Gingell, J. (1999). *Key Concepts in the Philosophy of Education.* London: Routledge.

Part I Changing perspectives of ecology

In this section Joseph Des Jardins provides an introduction on how to differentiate between the science of ecology and the social or ethical uses of "ecology." The chapter highlights how drawing ethical conclusions from scientific facts is neither simple nor direct. By providing a brief overview on the use of models and metaphors in ecology, Des Jardins introduces the reader to how interpretations of ecology can inform and misinform ethics. Jurek Kolasa and Steward Pickett remind us what ecology is as a science and share what practicing ecologists see as important science and social concepts to convey to educators about ecology. These authors provide a sense of how the discipline is changing and discuss implications for education and public knowledge of ecology. Kolasa and Pickett also examine how differing perspectives held by ecologists themselves contribute to gaps in ecological understanding between ecology, education, and society. David Slingsby and Susan Barker extend this discussion by examining the role and impact of learned societies, government agencies, NGOs, advocacy groups, media, schools, and educators in shaping public understanding. These authors conclude with thoughts on how ecological understanding can inform behavior.

JOSEPH R. DES JARDINS

2

Scientific ecology and ecological ethics: the challenges of drawing ethical conclusions from scientific facts

INTRODUCTION

A thing is right when it tends to preserve the integrity, stability, and beauty of the biotic community. It is wrong when it tends otherwise.

(Leopold 1949)

There is something deeply sensible about grounding ethical and political recommendations concerning the environment on scientific fact. Creating environmental policy that ignores the facts of environmental science would be absurd. If society expects schools to fulfill their mission in educating future citizens about the natural environment, it surely must require that students be well informed about science. Thus, it would seem that both environmental policy and environmental education should be grounded in environmental and ecological science.

But the connection between ecological science and normative conclusions is neither simple nor direct. Science tells us how the world *is*; the normative conclusions of ethics and policy tell us how the world *ought to be*. The quotation from Aldo Leopold that opens this chapter is typical of those who derive normative conclusions (the *oughts* and *shoulds*, *right* and *wrong* of environmental policy) from ecological facts. From at least the time of Leopold, many environmental activists have relied on ecological science to support their prescriptions. A wide range of philosophical perspectives, including Deep Ecology, ecofeminism, and social ecology, have explicitly tied their social, political, and ethical theories to scientific ecology. However, failure to distinguish clearly the realm of scientific fact from normative value can create significant problems. Not the least of these is the charge, often made by conservative critics, that schools are engaged in environmental and political advocacy, if not indoctrination, disguised as teaching science.

Thus, environmental educators are challenged to distinguish ecological science from ecological ethics, scientific education from political advocacy, scientific ecology from philosophical and political ecology. This chapter will review some of the problems facing efforts to connect environmental science with environmental policy and attempt to distinguish ecological science from those ethical and political perspectives also identified as "ecological."

We begin by reviewing some of the difficulties facing any attempt to connect the science of ecology with normative conclusions. Challenges come from two directions. First, a long tradition in philosophical ethics denies that any normative conclusions are entailed by facts. The philosophical view that ethical values are implied by natural facts is known as ethical naturalism. Critics charge that ethical naturalism involves an illegitimate inference from fact to value, or from *is* to *ought*. Indeed, this inference has been identified as the naturalistic fallacy and it is the most significant philosophical challenge to any attempt at grounding environmental policy solely in ecological facts.

Second, drawing normative conclusions from ecological science also faces difficulties in identifying exactly what constitutes an ecological fact. Competing, and perhaps incompatible, models and paradigms have characterized ecological science. Many terms used to describe ecological conditions – *community* and *health* are two that come to mind most immediately – are value-laden terms that, according to critics, are more metaphorical than factually descriptive.

PHILOSOPHICAL CHALLENGES: FACTS AND VALUES

The claim that one cannot justify evaluative judgments solely on the basis of fact is most commonly associated with the eighteenth-century Scottish philosopher David Hume. Hume, explicitly modeling his own method on the work of Newton, thought that careful empirical observation could distinguish between reasonable beliefs, what Hume called matters of fact, and "sophistry and illusion." His conclusions were that, while science could tell us all that we can know about nature, it couldn't provide the basis for drawing normative conclusions from nature. Empirical science could provide a rational basis for judgments about matters of fact, but there can be no scientific or empirical justification for normative judgments. Thus, here in Hume, we find the initial formulation of what has become canonized as the "is/ought" or "fact/value" gap.

According to Hume, empirical science can discover facts about what is or is not, but questions of value and oughts are to be found only by looking within one's "own breast" for a "feeling or sentiment" of approbation or disapprobation. It was exactly this subjectivist conclusion that motivated Immanuel Kant to attempt the rescue of rational ethics from the attack by empiricism. Kantian ethics, with its emphasis on moral duties and respect for individuals, emerged as the major modern alternative approach to ethics.

Kant rejected subjectivism and sought to maintain a rational basis for morality by an appeal to the autonomy of practical reason, or the rational will. According to Kant, ethical judgments can be rationally justified, but not by reference to their empirical or factual content. Moral judgments are rational only in terms of their logical form as universalizable (i.e., acceptable by any and all rational agents) judgments. Kant essentially accepts Hume's account that moral sentiments and feelings are subjective and therefore independent of rational assessment. But Kant denied that sentiments exhaust the moral domain. He also denied that empirical facts were the only possible justification for moral judgments. For Kant, judgments about what we ought or ought not to do are justified not by facts but because they would be assented to by any rational person. Both sides, however, continued to discount the relevance of empirical science in determining what is, or is not, ethically justified.

For much of the two hundred years since, moral philosophers in the West have seldom strayed far from the boundaries established by Hume and Kant. Those in the Humean tradition understand morality as a matter of personal sentiments or feelings. These are typically interpreted as a subjective, and therefore non-rational, basis for morality. Kantians agree that facts and values are distinct logical categories, but find another way to rationally justify ethical judgments. During these centuries, the challenges of the fact/value, is/ought gap (or the "naturalistic fallacy" in the twentieth-century language of G. E. Moore and later dubbed the "open question") became deeply rooted in Western philosophical ethics. No matter what the facts, the normative conclusion one should draw from those facts always remains an "open question." Empirical facts and ethical values are logically distinct.

Of course, within this tradition science continued to have a role to play in public policy and ethics and this is especially evident among the utilitarians. Utilitarians allow individuals to determine their own conception of value and attempt to bring rationality to their normative conclusions by prescribing the maximization of these individually chosen ends. Empirical science could prove very helpful in determining appropriate means for attaining value-laden goals. But science itself remains neutral concerning the value of the goals themselves. Thus, in the clearest example of utilitarian social policy, market economics relies on scientific methods to determine the appropriate means for attaining public ends. Within classical economics, individuals are left free to choose for themselves what they value; the role of scientific economics is to determine how to optimize the overall satisfaction of individual preferences.

A similar judgment can be made about how many utilitarians view the relevance of ecological science. A clear example of how environmental policy conclusions have been drawn from ecological science can be found in the writings of Gifford Pinchot, founder of the US Forest Service. Pinchot argued that scientific forestry is best situated for determining exactly which forestry policies would achieve the ethical goal of producing the greatest good for the greatest number. Consistent with the ethical tradition of Hume and Kant, science does not determine the content of public policy. Citizens are free to choose for themselves the uses to which they want to put forest products (in Pinchot's case, timber for building homes). Scientific forestry could determine the best means for "producing from the forest whatever it can yield for the service of man. The central idea of the forester is to promote and perpetuate its greatest use to men. His purpose is to make it serve the greatest good of the greatest number for the longest time" (Pinchot 1914). The legacy of Pinchot's utilitarian forestry remains alive today in the concept of multiple-use, which guides much of US Forest Service policy. Individual citizens are free to choose the uses to which they want to put forestlands; the Forest Service relies on ecological science to uncover the most appropriate means for attaining those ends.

ETHICS AND SCIENCE: FROM ARISTOTLE TO DARWIN

Philosophers were not always so unsure of the relevance of scientific fact to the content of normative judgments. Aristotle, in particular, thought that substantive ethical conclusions could be drawn from the facts of science. Indeed, Aristotelian science and ethics would be the philosophical view most compatible with attempts to connect scientific ecology with recommendations of social policy. It would not be too much of a stretch to claim that Aristotle's science and politics provided the first model for drawing ethical conclusions directly from biological facts.

Aristotle's science, like his ethics, was *teleological*. Aristotle's science differed in part from a more modern version in holding that we do not fully understand an object until we understand its *telos*, or characteristic or natural activity. Based upon his observations of nature, Aristotle believed that all natural objects do have a natural and distinctive activity. The goal of this activity, the fulfillment of its nature (sometimes called its purpose or function), was identified in Greek as the object's *telos*. Hence, we understand some object scientifically, according to Aristotle, only when we understand its natural activity or *telos*.

An example might help. Suppose that we seek a scientific explanation of a heart. In an Aristotelian view, we don't fully understand the heart until we understand its characteristic activity. A teleological explanation of the heart answers these questions: What is the distinctive activity of hearts? Or, what functions do hearts perform that distinguish them from other bodily organs? We have a complete scientific understanding of the heart only when we understand how it functions in the circulatory system. A teleological explanation would describe the heart as the organ that pumps blood throughout the body.

With this admittedly brief characterization, we can understand how the connection between scientific fact and ethical value could be made. All living things have a unique and natural characteristic activity. Through careful observation, we can discover the characteristic activity of each species. (Aristotle, of course, was the world's first great taxonomist, and the language of species and genus comes from him.) When we understand the characteristic activity of a heart, for example, we also come to understand what a *good* heart is. A good heart is one that performs its characteristic activity well, one that attains or fulfills its *telos*. A good heart pumps blood through the body in a regular, stable, and continuous manner and does so over a long period of time. The *telos* is the natural fact that bridges the gap between science and ethics. In this way, Aristotelian ethical judgments can be justified in terms of their factual content and ethics becomes as objective as science.

This teleological system was further developed in the writings of Thomas Aquinas in the thirteenth century, and it is perhaps through Aquinas's writings that these views have had the greatest influence on Western thought. Aquinas synthesized Christian theology and Aristotle's science and ethics. He interpreted the scientific and ethical teleology of Aristotle as evidence that a divine plan operates in nature. The characteristic activity of all natural objects results from God's plan.

Because God is assumed to be supremely good and because the activities and functions discovered in nature are God's purposes, the natural order can be equated with the moral order. Nature itself has a purpose, and the harmonious functioning of nature reveals the goodness of God's plan. From this perspective, "laws" of nature are interpreted both in the scientific sense of *descriptive regularities* that we discover in nature and, because these regularities are the result of the divine plan, as the *prescriptive norms* that we ethically ought to follow. Also from this perspective, the word "natural" carries both a descriptive (in contrast to artificial or human-made) and a normative meaning. To identify something as natural is to identify it as good and valuable.

Several aspects of this ethical tradition make it very appealing to modern environmentalism. The assumption that natural ecosystems are well ordered and harmonious, and that each part of an ecosystem, and especially that each species, has a distinctive role to play in the overall scheme, is very reminiscent of the natural law tradition. From this perspective, the way the world is (or would be if humans didn't interfere) is the way the world should be. Such reasoning is often used to prescribe a general policy of preservationism and non-intervention.

Unfortunately, several challenges to the Aristotelian tradition seriously undermine its relevance to contemporary environmental debates. The first objection denies that natural objects do have some one determinable and distinctive *telos*. Some objects do have a purpose or function. Human artifacts, such as a chair or a computer, are obvious examples. Some *parts* of a natural whole, such as a heart or a chromosome, also seem to have natural functions. But these parts have a function only in the sense that they contribute to the activity of the whole. It does not seem obvious that these wholes themselves have either a function or purpose. What, for example, is the characteristic activity of a human being? What is the characteristic activity of a spotted owl or a prairie? Philosophers and scientists in the modern era have thought that they could fully understand and explain natural objects without having to assume some natural purpose or plan. In fact, the scientific revolution of the sixteenth and seventeenth centuries was revolutionary in exactly the sense that it rejected the scientific relevance of teleological explanations.

A second objection denies that we can conclude that something is good simply because it is natural. Because some natural occurrences seem evil – Tennyson's nature "red in tooth and claw" comes to mind – some reason other than the appeal to nature must show why something is good. It would be a giant leap, for example, to reason from the characteristic natural activity of the HIV virus to the conclusion that this virus is good. In the natural law tradition, the explanation most often given for the connection between natural and good lies in a divine plan. However, this appeal effectively ends the philosophical discussion because it provides a reason only to those people who already assume a divine creator. To people of other religious traditions, or to people who do not believe in a supremely good creator, this reason carries little rational weight.

Finally, modern evolutionary science provides a significant, and perhaps insurmountable, challenge to this natural law tradition of connecting biological science with ethics. The process of natural selection offers an account of the apparent design found in nature without appealing to any purpose, or *telos*. The order that is found in nature comes not from some preordained plan but from the process of species adapting to their environment through random mutation and natural selection. For example, a defender of natural law might explain the long neck of the giraffe by claiming that the long neck exists (or was designed) in order to allow the giraffe to reach food high off the ground. Natural selection, on the other hand, claims that the giraffe did not develop a long neck in order to reach the leaves high in a tree but that it survived because, having a long neck, it could reach these leaves and was thus better adapted to its environment than competing organisms. The long neck functions to retrieve food from high places, but it was not designed to, nor is its purpose to do so. The neck is itself the result of random evolutionary change. Thus, the purposive language of teleology can be reduced to the more mechanistic language of the physical sciences. In this view, nature as we find it today is the result of hundreds of millions of years of random evolutionary change. Nature is neither good nor bad; it just is. The controversies that surround Darwin even to this day derive exactly from such implications.

To summarize, it is fair to say that by the early twentieth century a strong consensus had arisen among philosophers that normative judgments of morality and social policy could not be justified by appeal to natural facts alone. The role of science is to describe the world in a factual and objective way. The prescriptive domains of ethics and public policy are logically distinct from the descriptive world of science, allowing no direct inference from facts to values. The one Western tradition that had allowed such an inference, the natural law tradition of Aristotle and Aquinas, had been discredited because science no longer accepted the legitimacy of teleological explanations. Religious belief might assent to a purpose or plan in nature, but science could not. Those who seek to use the facts of nature to fully justify a normative judgment, without assuming an underlying religious framework, face formidable challenges indeed.

ECOLOGICAL CHALLENGES: MODELS AND METAPHORS IN ECOLOGY

A second set of challenges to deriving environmental prescriptions from ecological science comes from ambiguities within ecological science itself. The science of ecology arose in the intellectual context in which the old Aristotelian and Natural Law philosophies were in decline, but not yet absent. Ernst Haeckel is usually credited with coining the word "ecology" in the 1860s. Inspired by Darwin, early ecologists were committed to the study of organisms in their natural environment. The fruitfulness of this approach soon paid off as ecologists came to understand that the biological world was more complex than the mechanical models of the physical sciences would suggest. Unlike botany and zoology, which tended to focus on discrete plants and animals, ecology emphasized the connections that exist in nature. Individual plants and animals are connected to other individuals in their own species, to individuals of other species, and to the physical environment in which they live.

But the nature of this interconnectedness is open to various interpretations, some of which explicitly renew the alleged link between science and ethics. Some of the earliest pioneers of ecology conceived of the relationship between individuals and their living and nonliving environment on an organic model. Henry Cowles and Frederick Clements, for example, separately studied plant succession and concluded that for any given location and climate, vegetation develops toward a relatively stable and permanent grouping, what Clements called the "climax community." Clements explicitly thought of this community itself as a "complex organism." On this account, ecosystems go through a normal and natural developmental process that builds toward a final natural equilibrium. Just as any single organism grows through developmental stages toward a mature level, so, too, do ecological organisms grow, develop, and mature. Ecological communities can therefore be described as healthy, diseased, young, mature, and the like, according to a normal and natural developmental standard.

Reminiscent of the natural law tradition, the organic model would seem to provide a scientific basis for identifying and diagnosing problems and offering advice for solving environmental problems.

According to the organic model, ecosystems strive toward a natural healthy condition, a stable and unified state of balance and harmony. To the degree that ecological communities do have a unified, stable, and harmonious end toward which they are developing - that they do have a *telos* discoverable by science - then reaching normative conclusions on the basis of scientific facts once again seems possible. Just as with the heart example, the value of individual species can be judged in terms of how they contribute to the well-being of the organic whole and the condition of the organic whole can be judged in terms of overall health.

Given the religious and cultural controversies that surrounded Darwin, it would not be too much of a stretch to see this organic model in ecology as a resurrection of the idea of a natural order and design in nature. Indeed, there has been some suggestion that early ecologists hoped to do exactly this. What physics and astronomy took away in the sixteenth century, ecology could bring back in the nineteenth (Chase 1995). However, two general challenges work to undermine the legitimacy of the organic model as a basis for drawing evaluative conclusions.

First, the facts began to overwhelm it. Ecological communities do not always develop toward one single and stable climax community. The type of unity and stability observed by Cowles and Clements might be true in specific locations over the short term, but it is not true in general and over the long term. The facts observed by such early ecologists as Cowles and Clements were severely restricted by time and place. What is observed as stability at a specific location over the short term turns out to be unstable over the long term.

Second, some ecologists find the language of superorganisms and organic wholes unscientific and metaphorical. The organic framework lends itself to a descriptive language at home in medicine and health, but when applied to natural environments it seems more the result of philosophical presuppositions than empirical observation. To this day, the concept of ecosystem health, a clear echo of the organic model, remains a highly contentious scientific notion (Constanza *et al.* 1992).

Influenced by the more positivistic approach of the physical sciences, alternative models for understanding ecological interconnections developed in the early decades of the twentieth century. British ecologist Arthur Tansley defended the concept of an ecosystem as a more appropriate model for ecological research. Ecosystems were to be understood as a complex of related physical factors in much the same way that the solar system or a single atom is a complex of physical parts. Tansley explicitly sought to purge ecology of the metaphorical, and value-laden, language of the organic model and replace it with what he saw as the more descriptive and value-free language of the physical sciences (Golley 1993).

If ecosystems are nothing more than physical systems, and the interconnections among them were describable in physical terms, then the gap between scientific fact and ethical value would remain unbridgeable. It would make no more sense to conclude that an ecosystem *should* be arranged the way it is than to conclude that the planets *should* revolve around the sun at the distance and speed at which they do.

Beginning with Tansley, ecologists focused more on the structure and function of ecosystems. A system's structure refers to how the many parts are related. The function of any individual part is explained in terms of how it contributes to the activity of the whole. A key concept introduced with the ecosystem idea was the notion of a "feedback loop." Elements within an ecosystem are related not simply in linear and causal ways but in more complex ways characterized as feedback loops. Essentially, this means that the elements within a system not only are affected by other elements, but they in turn produce effects upon other elements in a dynamic network of interconnections. This feedback is not random but works to maintain a balance or equilibrium within the entire system. The standard example of a feedback loop is the thermostat on a home heating system. Falling room temperature triggers a reaction by the thermostat, which in turn changes the room temperature by turning on the heating system. The change in room temperature affects the thermostat, which shuts off the heating system until, yet again, room temperature falls and the cycle begins anew.

Thus, this approach holds that nature is organized into ecosystems–grasslands, lakes, prairies, forests – that are structured in such a way that, through the normal functioning of the individual members, the systems maintain a relatively stable equilibrium, much like a heating system maintains room temperature. Both the structure – the network of relationships – and the function – the activities of feedback loops working to maintain equilibrium – can be explained in scientific and mathematical terms. Thus, the holism of ecology is incorporated in the prevailing physical scientific paradigm.

One of the most influential and respected scientists associated with the community model was the English zoologist Charles Elton. Elton described his zoology as "the sociology and economics of animals." His goal was to present a description of nature as an integrated and mutually dependent economy. Elton's community model therefore was a *functional* model: individual members are identified by the food function that they perform in the system. The system is seen in economic terms. Some members function as producers, some as consumers. The commodity is food, and ecological communities can be described as "food chains" in which individual members fill various occupations. The laws of ecology thus describe the processes of producing, distributing, and consuming food. Accordingly, an individual species' function or role within a food chain – its ecological *niche,* in Elton's terms – is determined by what it eats and what eats it.

The idea of a food chain is perhaps the most familiar concept of the community model. Some organisms, called "producers," manufacture their food by producing organic compounds (sugars, starches, cellulose) from inorganic molecules (carbon dioxide and water) and energy. Photosynthesis is the primary process through which producers manufacture food. Other organisms, "consumers" (herbivores, carnivores, omnivores), depend upon producers, directly or indirectly, for their food source. At the far end of the food chain, decomposers (mostly fungi and bacteria) feed on dead organic material, breaking it down into simpler inorganic molecules. Producers can again use the inorganic molecules, whereas worms, insects, and other organisms in turn eat the decomposers.

Although this community model continues to influence both ecologists and others, like the organic model it also faces criticism for being overly metaphorical and unscientific. "Community," "household," "producers," and "consumers," are, at least, ambiguous when applied to non-human nature. When used in a human context, each of these terms is intentional; that is, each implies a purpose, goal, or intention that is the result of conscious choices. When applied to a natural ecosystem, they are functional but not intentional. Not since the abandonment of the teleology of the natural law tradition has science suggested that nature has a purpose. Presumably, no scientific ecologist claims that plants produce organic matter *in order that* herbivores have something to eat, or that bacteria decompose an animal carcass *in order to* return inorganic material into the ecosystem.

However, it would seem that it is the intentional interpretation that is needed to draw any normative conclusions from scientific facts. As in the case of the giraffe's long neck, we must distinguish functions from purposes. Functions can be the result of prior physical causes (e.g., natural selection); purposes and intentions result from choices.

The significance of this fact is that attaining a purpose is (in some sense) good in a way that attaining a function is neither good nor bad. Purposes are chosen and that means that they are good at least in the sense that the chooser perceives them as good. Functions imply no such choice and, hence, no such (perceived) good. Even if plants fill a function within an ecosystem, it remains an open question if this fact is good or bad.

We might understand this point better by returning to the thermostat example. One cannot have a complete understanding of a thermostat without knowing how it functions within a heating and cooling system. But this does not imply, at least in so far as science or engineering is concerned, that the thermostat itself has any intrinsic purpose, design, or intention. The thermostat is simply a mechanical system that operates according to causal laws. Its purpose comes from the external designer and it is a good thermostat only to the degree that and only for so long as it accomplishes its designer's purpose. But ecology cannot make a similar claim about the natural functions found within ecosystems. Whose purpose, design, or intention would it be? This means that it is not a matter for science to decide if an organism serving its function is good or bad, right or wrong.

Within the systems approach in ecology, the community model continues to compete with a less metaphorical and anthropomorphic version. Consistent with Tansley's desire to legitimize ecology by connecting it with the physical sciences, some ecologists deemphasize such qualitative terms as food, producers, consumers, communities, and occupations and replace them with the seemingly more objective language of ecosystems and energy. In this *energy model,* the focus of ecological research is on the ecosystem as an energy system or circuit. Just as the physicist studies the flow of energy through a physical system, the ecologist studies the flow of energy through an ecosystem. The language of a food chain is replaced with the mathematically more precise language of chemistry and physics. Like the thermostat, an ecosystem appears as just another physical, mechanical system.

We can trace the flow of energy through an ecosystem in a way that parallels the flow of food through the food chain but without using economic or household metaphors. Photosynthesis is the process through which solar energy breaks the chemical bonds of carbon dioxide and water molecules, forming new molecules of carbohydrates and oxygen. Respiration transforms carbohydrates and oxygen back into carbon dioxide, water, and energy. The energy released in this process powers the chemical and physical processes of life, growth, reproduction, and so on. Photosynthesis and respiration are the principles of the carbon and oxygen cycles in ecosystems. A similar account can be given for the nitrogen cycle.

Because the carbon, oxygen, and nitrogen cycles – as well as a similar cycle for phosphorus and the more familiar water cycle – are all ultimately driven by solar energy, ecologists can account for ecosystems in terms of the energy that flows through various chemical, biological, and climatic cycles. The energy model strives for, and perhaps attains, a more objective and less value-laden interpretation of ecosystems. The fact that carbon or nitrogen cycles through a process from non-living to living to non-living things is one thing; the judgment that this is good or bad, right or wrong, is logically another.

For the most part, each of the ecological models mentioned to this point share a common assumption. Natural ecosystems tend toward a point of relative stability or equilibrium. When the system is disturbed, natural forces work to return to a point of equilibrium; a system in equilibrium tends to resist change and stay in equilibrium. Aldo Leopold, writing in the 1940s, was influenced by the organic, community, and energy models and therefore could conclude that all ecosystems did move toward their own point of integrity and stability.

More recently, however, many ecologists have challenged even this conclusion, arguing instead that natural systems are more non-equilibrium than previously thought (Botkin 1990). This view claims that ecosystems are constantly changing and, perhaps more importantly, that this change occurs without direction or any sense of "development." Disturbance resulting from other physical and ecological processes are so frequent that no biotic community remains stable or in equilibrium for long periods. The non-equilibrium model denies that any one arrangement can be uniquely characterized as *the* natural structure, identity, or equilibrium of an ecosystem. There is no normal balance or long-term equilibrium that exists within natural systems.

FROM SCIENCE TO ETHICS AND BACK AGAIN

Ecology is proving to be very fertile ground for scientific theorizing and modeling. It may be fair to characterize ecology as still a young science that is yet to find its dominant paradigm. The organic and community models encourage teleological interpretations of nature that would allow an inference from scientific fact to ethical value. Yet their crucial teleological elements – community, organism, health, producers, and consumers – appear, at bottom, metaphorical. The less metaphorical and more mathematical models of energy flow and non-equilibrium model seem to offer little direct support for normative conclusions. Perhaps the significance of this gap between ecological science and ethics can best be seen when we consider the range of implications that might be drawn from these various ecological models.

Consider any ecological theory which holds that ecosystems develop toward a natural equilibrium or stability. For many environmentalists, this fact would imply a policy of preservation or protection. Left alone, nature will find its own best way to stability, balance, and harmony. The natural harmony and cooperation within ecosystems guide us toward a policy of respect for nature's way and the preservation of natural systems. There is a natural balance in nature and we ought to respect it.

But consider an alternative argument. Why not conclude that, because nature does work toward balance, humans can take a much more active and involved role within nature? Because the forces of nature work toward a harmonious equilibrium, humans need not be overly concerned about upsetting that balance. Further, because we can discover the mechanisms through which this balance is maintained, we are in an even better position to manage and control the ecosystem. We can learn to create and manage the harmony of natural processes, perhaps by adding nutrients and fertilizers to the soil, perhaps by killing non-native species, perhaps by introducing new predators or genetically modifying a species to help it adapt to a changing environment. Thus, the equilibrium thesis might be used to defend policies that tread both lightly and forcefully.

Similar implications could be drawn from a more non-equilibrium model of ecosystems. Imagine that we do conclude that the governing principle of ecology is change rather than stability, and that the natural state of things is in flux rather than harmony and equilibrium. What follows? Some could use this as evidence to argue against preservation and restoration because there is no natural order to preserve or restore. Paralleling the Social Darwinists of the nineteenth century, non-equilibrium theorists might argue that, because all species struggle to survive in a constantly changing world, humans also have a natural inclination to manage their environment for their own self-interest. On the other hand, some might just as well argue that the very non-equilibrium nature of ecosystems means that we should be even more cautious of our activities rather than less. This complexity is more, rather than less, of a reason to stand in awe of, and respect, the natural world and to be much more modest about our understanding of, and ability to manage, nature.

The point of this exercise is to demonstrate that we can draw diverse ethical conclusions from even the most uncontroversial facts of science. Given this, it is perhaps not surprising that environmental education is open to criticism on political grounds, especially from conservative critics. If scientific facts are not sufficient to establish policy conclusions, then any policy conclusions found in environmental education must, or so it seems, come not from science but from the teacher's own political agenda.

Thus, despite the importance of ecology to environmental concerns, this science alone is not sufficient to decide among competing environmental policies. No matter what the facts, normative implications remain an open question. If one does wish to ground environmental policy in ecological science, further normative premises will need to be provided to connect scientific fact to policy conclusion. We now turn to a consideration of some of the normative conclusions that might be grounded in the science of ecology.

THE SOCIAL AND ETHICAL USES OF ECOLOGY

It is an understatement to say that the word "ecology" and the corresponding prefix "eco" have become ubiquitous outside of the biological sciences. One can find ecology modified by such adjectives as "deep," "shallow," "social," "radical," "restorative," "human," "Marxist," "literary," and "urban." The prefix "eco-" or the term "ecological" can regularly be found modifying "feminism," "spirituality," "justice," "literacy," "conscience," "ethics," "economics," and "psychology." With just a little work, one could find countless other uses of "ecology" outside of the biological sciences.

Some of these uses seek to make a direct connection between normative judgments and scientific ecology. Aldo Leopold and many interpreters of his land ethic would fit this model, as would many defenders of Deep Ecology. Others make the connection less direct. "Urban ecology," for example, studies humans in their urban setting and seems to use "ecology" to emphasize the fact that humans are being studied within a particular urban environment. Literary ecology and ecocriticism study the relationship between literature and the natural environment. But even in these less direct cases, one suspects an underlying agenda. Identifying an approach as "ecological" suggests that we will find such interconnections and that these will be intellectually meaningful and valuable. The fruitfulness of scientific ecology seems to testify to the validity of both assumptions.

But in what sense will such interconnections be meaningful and valuable? Clearly, for some, the interconnections imply a range of normative conclusions. *Because* the world is this way, we *should* act in these ways. We have seen the two most important challenges to those who use scientific ecology and ecological concepts to justify political, social, and ethical conclusions. First, there is a logical gap between facts and values that makes any immediate inference from the science of ecology to normative conclusions suspect. No social or political position follows logically from any ecological fact. Second, one must be particularly careful when appealing to some ecological concept or conclusion in normative debates. Within scientific ecology many of the most normatively useful terms – integrity, health, community, balance, stability – are often the most scientifically contentious. One risks basing one's normative conclusion on some ecological "fact" that ecologists themselves would not accept.

These challenges have led most thoughtful environmentalists to be hesitant when appealing to ecology. For example, Arne Naess, the Norwegian philosopher who introduced the distinction between shallow and deep ecology in 1973, is explicit in rejecting a direct reliance on scientific ecology for normative conclusions. Naess believed that it is a mistake to think that scientific ecology alone can resolve environmental disputes. Rather, scientific ecology can provide models and metaphors that can be useful in changing the way humans think about and value the natural world. Only when such a change in "worldview" occurs can we hope to fully address the ecological crisis (Naess 1989).

In this final section I would like to follow Naess's advice and offer some tentative suggestions for how scientific ecology might guide our thinking about matters ethical and political. To help in this process, let us use an example of a proposed environmental policy: reintroducing woodland caribou and elk into the Boundary Waters Canoe Area in northern Minnesota. How might ecological science contribute to this policy debate?

First, as suggested throughout this chapter, no direct policy conclusions are entailed by any scientific facts. One cannot argue, for example, that because caribou and elk were part of the natural ecosystem and were native to the region prior to widespread human hunting and logging, they should be reintroduced as part of the natural ecosystem. More is needed beyond the simple facts of science. "Natural" and "native" here are warning signals. Neither was particularly natural or native when a deep ice pack covered the area during the last ice age. Nevertheless, science does have roles to play in this discussion.

Philosophers often cite a traditional adage that "ought implies can." It is unreasonable to argue that one ought to do what, in fact, one cannot do. Thus, one role for ecological science is to establish the parameters of what can be done. Ecology can determine the food and habitat requirements for caribou, for example. If the Boundary Waters Canoe Area lacks an adequate supply of food or appropriate habitat, then a policy of reintroducing caribou or elk would be scientifically inadvisable.

Ecological science can also guide policy by describing the likely implications of policy decisions and evaluating possible means to alternative ends. Ecological science could describe the expected effects that a reintroduction would have on deer, moose, wolf, and plant populations. Science could tell us the likelihood of caribou or elk survival without proper food. Science could identify alternative sites for the reintroduction, or suggest means for insuring an adequate food supply.

Setting the parameters for what can be done and identifying the likely consequences of various alternative policies are two important ways in which environmental policy should be grounded in ecological science. But ecological science can make some more general contributions as well. Let me offer three general and closely related lessons that environmental policy should learn from scientific ecology.

First, scientific ecology should teach us to think of the world as interconnected with complex feedbacks. There are patterns and interconnections in the natural world that are easily missed and underestimated. Reintroducing caribou and elk should not be decided in isolation of many other factors, including its effects on populations of other species, long-term likelihood of success, and so forth.

Second, scientific ecology teaches us that radical anthropocentrism is misguided. Human beings are not independent of, nor are they particularly unique within, the natural world. The causal relations between species run both ways. Motorized recreation with snowmobiles and ATVs, logging, and especially loggings' roads can have significant impact on caribou and elk populations. But, as recent cases of outbreaks of chronic wasting disease, foot and mouth disease, and monkey pox demonstrate, humans are well advised on self-interested grounds to pay close attention to the diseases in animal populations.

Finally, uncertainty itself is a valuable lesson to be learned from ecology. Ecological events are part of a complex and non-linear system of relationships. Ecological facts are most appropriately stated in probabilistic and tentative terms. As a result, scientific ecology teaches us that a particularized, contextual, and cautious examination is often more fruitful for understanding than an overarching and generalized theory. A reasonable implication of this is that environmental policy should proceed cautiously.

It seems to me that in their own way each of these lessons suggests a policy of caution and prudence in our dealings with natural ecosystems. Whether our environmental values are aesthetic, moral, spiritual, or self-interested, we are well advised to go slow when we change, manipulate, and use natural ecosystems. The complexity of natural ecosystems and the mutual dependence that exists between and among species and between and among biotic and abiotic components of the biosphere strongly argue against certain policies and practices. Policies that are irreversible, that disrupt ecosystems significantly, and rely on human and technological back-up and safety measures should face a major burden of proof before being approved. Scientific ecology does not prove that policies for nuclear waste disposal, genetic modification of crops, releasing carbon dioxide into the atmosphere, alteration of rivers and other waterways, draining wetlands, modification of soil composition, clear-cutting old-growth forests, or which threaten endangered species should be abandoned. But science does provide us with good reasons for approaching such decisions with humility rather than arrogance.

Ecological science also cautions us against the assumption that concepts and categories that work in one time and place will work in all times and places. Humans have the ability to manipulate and dramatically change natural ecosystems. Despite our best intentions, there can be no guarantee that such changes, on either the local or global level, will continue to support human interests. Scientific ecology warns us that, if we ignore the interconnections and dependencies of the natural world and continue to seek a dominance of the natural world, we do so at our own peril.

REFERENCES

Botkin, D. (1990). *Discordant Harmonies*. New York, NY: Oxford University Press.

Chase, A. (1995). *In Dark Wood: The Fight over Forests and the Rising Tyranny of Ecology*. Boston, MA: Houghton Mifflin.

Costanza, R., Norton, B., and Haskell, B. (1992). *Ecosystem Health*. Washington, DC: Island Press.

Golley, F. (1993). *A History of the Ecosystem Concept in Ecology*. New Haven, CT: Yale University Press.

Leopold, A. (1949). *A Sand County Almanac*. New York, NY: Oxford University Press.

Naess, A. (1989). *Ecology, Community, and Lifestyle,* ed. D. Rothenberg. Cambridge: Cambridge University Press.
Pinchot, G. (1914). *The Training of A Forester.* Philadelphia, PA: J. B. Lippincott Co.

FURTHER READING

Callicott, J. B. (1989). *Defense of the Land Ethic.* Albany, NY: State University of New York Press.
Des Jardins, J. (2001). *Environmental Ethics: An Introduction to Environmental Philosophy.* Belmont, CA: Wadsworth Publishing.
Norton, B. (1991). *Toward Unity Among Environmentalists.* New York, NY: Oxford University Press.
Rolston, H. (1986). *Philosophy Gone Wild.* New York, NY: Prometheus Books.
(1988). *Duties to and Values in the Natural World.* Philadelphia, PA: Temple University Press.
Weston, A. (1999). *An Invitation to Environmental Philosophy.* New York, NY: Oxford University Press.
Worster, D. (1985). *Nature's Economy.* Cambridge: Cambridge University Press.
(1993). *The Wealth of Nature.* New York, NY: Oxford University Press.

JUREK KOLASA AND STEWARD T. A. PICKETT

3

Changing academic perspectives of ecology: a view from within

INTRODUCTION

Ecology is a science. This statement may surprise some readers who will see it as superfluous and stating the obvious. However, because the public, at whom environmental education is aimed, may see ecology as a cause or a political agenda, being more specific about what ecology is and why ecology is a science may be useful. This chapter is our personal answer to the question of what practicing ecologists would like to share with environmental educators about their discipline or, more specifically, what ecologists see as important to convey to others about ecology. The scope of the answer is restricted to a discussion of two aspects of ecology that have direct implications for environmental education and public knowledge of ecology: the *changing content and methods* of ecology and *changing social aspects* of ecology.

The first aspect focuses on the changing paradigms, conceptual constructs, and methodological approaches in ecology. The second aspect highlights changing views held by ecologists on the state of their discipline and public perceptions of ecology, and science in general. These two aspects, while strongly interacting, offer distinct challenges and opportunities when it comes to education and conveying what ecology is and its role as a science to non-ecologists.

WHAT IS ECOLOGY?

Ecology is the scientific discipline that investigates interactions of organisms with each other and their environment, addressing questions about processes that influence distribution and abundance of organisms, and the fluxes of matter and energy through the living world (Pickett *et al.* 1994). This includes the study of groups of organisms,

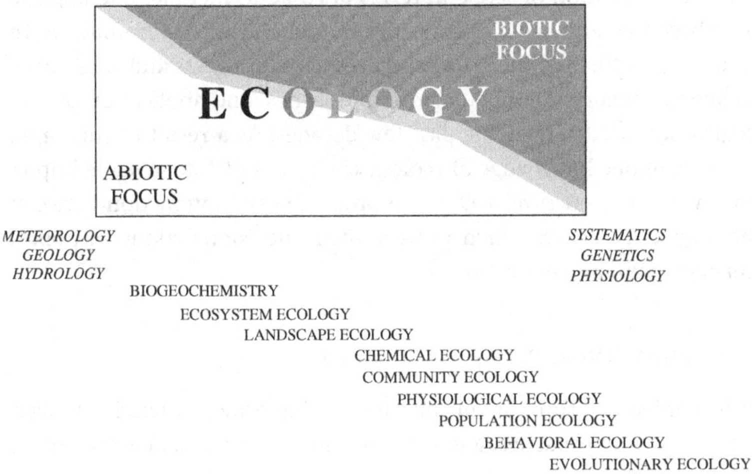

Figure 3.1. Gradient of ecological research areas arranged from those with primary focus on abiotic to biotic aspects of environment. This gradient changes as some areas discover links and interactions (from Pickett *et al.* 1994).

such as bird breeding pairs, wolf packs, symbiotic species such as lichens or corals, or whole populations and groups of populations living together in communities. Interactions among organisms and their environment include transformations of local microclimate, formation of soils, nutrient cycling, retention of water and soil, as well as large-scale phenomena such as the role of vegetation and its consumers in absorbing, storing, and releasing oxygen, carbon dioxide, or nitrogen.

Since ecology involves the study of diverse ecological phenomena at various spatial and temporal scales, the discipline has become specialized across a broad range of sub-disciplines (Figure 3.1). The sub-disciplines can be organized along a topic gradient from physical (abiotic) to biological (biotic) phenomena. The scope of ecology also intersects with many other physical and biological sciences, such as geology and genetics. In addition, ecology interacts with conservation, land and wildlife management, agricultural, land and water restoration, epidemiological and other related management, human health, and social sciences (e.g., McDonnell and Pickett 1993; Pickett *et al.* 1998).

Reflecting these interests, the scope of ecology is quite diverse in terms of phenomena investigated, interdisciplinary interactions, and overlap between fundamental and applied sciences. Therefore, ecology's precise delineation of scope in terms of content, method, and application becomes somewhat difficult to comprehend and communicate. To add to the difficulty, there has also been rapid progress and substantial change in theory, methodological approaches, and application of ecological knowledge over the past few decades. As a result of this rapid change, public knowledge of ecology can be out of date in some important ways (Pickett *et al.* 1994), compromising the interaction between ecology and society, which in turn affects decision-making in policy, management, and education.

CHANGING CONTENT OF ECOLOGY

It is emphasized throughout this chapter that ecology includes a diversity of areas, approaches, and specializations. At first glance, one might see few connections and shared themes among them. Nevertheless, closer analysis reveals the existence of a number of distinct and complementary paradigms. A paradigm consists of the set of background assumptions that a discipline makes. Another way to summarize the idea of paradigm is that it is the worldview that the scientists in a discipline hold. Paradigms mold subject area, approaches, and modes of problem solving (Kuhn 1962). Criteria of observation, the perspectives taken, and kinds of processes involved, and kinds of interactions included (Allen and Hoekstra 1992) are often different between paradigms. The following account of paradigms in ecology is adapted with modifications from Pickett *et al.* (1994).

Recently, ecology appears to have been structured around two major paradigms: one that represents population ecology, inclusive of related disciplines, and another that represents ecosystem ecology. The primary distinction between these paradigms is their focus on organisms on the one hand and on materials and energy flow on the other. Such a distinction is clearly apparent in ecological, environmental science, and biology textbooks that differ in their coverage and focus.

Population paradigm

In the population paradigm, emphasis is on the individual organism, or populations and communities composed of individual organisms, as the basic units of study (e.g., Begon *et al.* 1990). The population ecology paradigm addresses patterns and causes of change in the distribution and abundance of organisms in space and time (Silvertown 1982). Such a broad perspective comprises a range of different concepts. It may include evolutionary constraints in mating systems with, for example, questions of how parental investment determines population success in environments with fluctuating food resources. It may also examine allometric rules where species abundance is explained with the aid of body size and metabolism. Finally, gradients of diversity at the macroecological scales of regions and continents may be compared with each other to tease out major determinants of diversity.

Abiotic factors are usually considered to be external forcing functions that control the life history of species. By affecting life histories, abiotic factors alter the dynamics of organisms and aggregations of organisms. In most studies, the ecosystem containing the population is taken as a prerequisite that does not require explicit treatment. For example, studies of niche partitioning, or the division of resources such as habitat or food items, among Lake Victoria cichlid fishes do not concern themselves with the nutrient budget of the lake (Leveque 1995). Thus, to answer the questions commonly posed by population ecologists, changes in material and energy flows are not usually well connected to organismal dynamics or historical processes constraining those dynamics (but see Robertson 1991).

Ecosystem paradigm

The common view of the ecosystem paradigm focuses on patterns of material and energy flow and the processes controlling them (e.g., Reiners 1986; Schlesinger 1991). The abiotic component of the environment is explicitly included in the ecosystem. At the same time, the complexity of interactions among species, between species and their environment, and spatial heterogeneity of organisms and processes are often "black-boxed" and taken as constants (Damuth 1987; Cale 1988).

Packing organismal characteristics away in an opaque box in ecosystem ecology does not mean that they have no intrinsic interest or significance in other processes. Rather, this perspective is one approach to necessary simplification in ecosystem ecology, permissible because considerable progress toward the main goal of understanding ecosystem structure, function, and change can be achieved primarily by measuring material inputs, outputs, transformations, and pools. Both the abiotic components and organisms in ecosystems regulate fluxes. However, in most studies, organisms are aggregated on the basis of their role in processes of flux. The categories of producers, decomposers, N-fixers, and shredders are examples of the aggregations of organisms conceived in ecosystem studies. How detailed organism identities and dynamics influence fluxes is not usually central (but see Marks 1974; Vitousek 1989).

Like other areas of ecology, this perspective continues to evolve. Some ecosystem theorists believe that the black-box approach is not a complete ecosystem perspective (e.g., O'Neill *et al.* 1986) and that the ecosystem concept itself needs major revision (O'Neill 2001). Debates about the nature of the ecosystem concept are not unique to this paradigm - the population concept is continuously revised as well (e.g., Baguette and Stevens 2003).

Since the foci, scales, and criteria of observation in population and ecosystem ecology are usually so different, their paradigms and constituent assumptions, approaches, and lexicons are also different. These differences are rational and helpful for focusing research. No doubt major advances in ecological understanding will continue in each of these sub-disciplines. However, without understanding these differences, concepts from each paradigm sometimes become conflated, or one paradigm dominates the other in education or management of resources.

Integration of population and ecosystem paradigms

In the broadest sense, ecology is now understood to be "the scientific study of the processes influencing the distribution and abundance of organisms, the interaction among organisms, and the interaction between organisms and the transformation and flux of energy and matter" (Likens 1992). From the perspective of this broad definition, the integration of the population and ecosystem sub-disciplines has become one of many potential goals for advancing ecological understanding. Such a goal can be achieved by forging links across these sub-disciplines and by focusing on ecological issues and critical questions that lie at the intersection of the sub-disciplines (Jones and Lawton 1994) and, therefore, cannot be addressed by either of the two paradigms already described but requires input from both paradigms.

Other paradigms

The two paradigms described above are not the only ones ecologists have identified. Another paradigm, of particular interest to management and environmental education, is the "balance-of-nature" worldview. Wu and Loucks (1995) point out that the "balance-of-nature" view once dominated and motivated ecological research, primarily in the 1960s and 1970s

(Egerton 1973). The balance-of-nature, or equilibrium, view affected and still does affect choice of research question, interpretation (Cuddington 2001), and application of results. For example, phenomena such as historical effects, spatial heterogeneity, and stochastic factors, including disturbance, have been assumed to play a minor role in the equilibrium view. Furthermore, biodiversity has been seen as a carrier and agent of stability.

A more recent framework of non-equilibrium ecology has largely supplanted this paradigm (Pickett *et al.* 1992). In this framework, phenomena, patterns, and relationships are viewed as arising and maintained by disruptive, external stochastic agents such as tree falls, fires, transient grazing, or rock and ice floe scouring, which overlay local processes of competition, predation, and mutualism. For example, a sequence of changes in populations and community structure in a habitat may be the result of regional dynamics or metacommunity dynamics (e.g., Hanski 1997).

This equilibrium–non-equilibrium duet of paradigms permeates the two primary paradigms, population and ecosystem worldviews, adding to the difficulty of communicating ecological findings and principles. The task of communication is compounded by the fact that all these paradigms are "works in progress", whether of development or abandonment, and do not always express themselves in a pure form. Indeed, it is a characteristic of paradigms that they are rarely articulated, but rather operate in the background of the scientific process. The paradigms presented above also represent contrasts and endpoints rather than a complete taxonomy of ideas. Consequently, many specific research activities, agendas, and conceptual constructs may contain flavors of two or more of the paradigms and will not easily fall into one category.

Implications for education and public knowledge of ecology

The public appears to hold on to the balance-of-nature view, whereas ecologists have moved on to a much richer, non-equilibrium perspective on the relationship between organisms and their environment played out at on multitude of spatial and temporal scales. The public perception is particularly evident within environmental movements and writings. This perspective has, until recently, also dominated in some areas of management and led to major undesirable consequences.

For example, Yellowstone National Park has been "protected" from the disrupting effects of fire in the erroneous belief that disturbance disrupts the harmony and well-being of its biotic components. For many years fires were controlled and suppressed because of this view. It turned out, however, that fire is a defining factor of habitat heterogeneity and necessary for persistence of large-scale patterns of forest stands and animal diversity. A narrow view of balance-of-nature, or equilibrium, failed as guidance to management. The equilibrium view had to be replaced by one in which changes on the fine scale and in the shorter term reflect individual fires and succession in patches after fire contribute to a larger mosaic of patches in the landscape. Although the landscape mosaic has not been stable over the very long term, it is resilient because a mix of burned, non-burned, and recovering patches has continued to exist within it (Romme and Knight 1982).

The notion of equilibrium itself turns out to be multifaceted. On one hand it entails beliefs that natural populations fluctuate around constant means, that communities have a more or less constant number of species, and that communities of species maintain a "delicate balance" of relationships (Cuddington 2001). However, these claims are true only for some of the time and in some places. The perception of equilibrium often depends on the time and spatial scale, and because ecological systems continuously change and develop, ecology does not support the balance-of-nature metaphor. Unfortunately though, ecologists often use this metaphor in spite of knowing well that research does not support it.

Environmental groups and the media can amplify the misuse of these metaphors. Amplification of views that have little or no scientific basis can lead to unproductive or, worse, counterproductive arguments about how to best manage natural living resources. In simple terms, one cannot draw a boundary around a piece of nature and insist that, if left alone, this piece will harmoniously persist for our admiration and benefit. The size of the system, its openness, time scale of interest, and its components will all determine the degree to which it can self-regulate, retain the ability to adjust to large-scale natural phenomena such as weather, and to what degree we need to intervene to retain its essential features. There is simply no science-based catchall answer to this most common of issues in our relationship with nature.

THE CHANGING SCIENCE OF ECOLOGY

One criterion of science is to have laws, but this criterion is not necessarily a defining one. The mere use of and reliance on laws, even if formulated outside ecology, make it a science (Mahner and Bunge 1997). Laws imported from other sciences include the law of natural selection from evolution (Mayr 1982) or laws of conservation of matter and of stoichiometry from chemistry (Sterner and Elser 2002). Whether or not ecology can reach a consensus of what its laws are may not be crucial at this stage. Laws in ecology can have the form of conditional statements, indicating if some condition holds, then some result will occur. Even such conditional laws can be probabilistic, or statistical. Other laws take the form of generalizations inductively derived from many cases observed over time and space. What is important is that ecology does rely on and search for laws. Ecologists have cast such processes as succession (Pickett and McDonnell 1989) as laws. Boundary function may also have a similar law-like formulation (Cadenasso *et al.* 2003).

In this search for laws, ecology employs a rich array of approaches and methods, some of which may be quite complex blends of simpler tools. A few general categories stand out and have acquired considerable influence. These categories include experimentation, comparisons, mathematical models, and pattern analysis. Each of the categories can address an array of ecological systems and scales. For example, experimentation spans a broad range of ecological systems, from microcosms (e.g., McGrady-Steed *et al.* 1997) to whole lakes (e.g., Carpenter 1996). At the same time, questions asked by experimenters depend on which ecological sub-discipline is of interest and may involve such distinct areas as bee behavior and landscape soil erosion. Distinct systems, scales, and questions of interest allow for many specific combinations that, irrespective of their uniqueness and diversity, share the common attention to strictest criteria of scientific inference.

This diversity, if sampled incompletely or superficially, may lead to an impression that ecology provides unsound conclusions and unreliable advice. This is a false impression. The diversity and debates are evidence of strength and growth but, like any developing system of knowledge, must be taken with caution and with the assistance of expert advice. Below is a sketch of a few of the most active and diverse areas of discussion regarding ecological research that might puzzle and confuse an educator with a non-scientific background. This is even more important if one notes that individual ecologists differ in the emphasis they place on particular paradigms, approaches, scales, methods, or results of research. Disagreements may sometimes be confusing to ecologists themselves, reflecting the social aspect of science, even when the results have an objective and sensible interpretation.

Approaches

An overview of ecology recognizes two distinct competing traditions or approaches to research methods. One tradition is reductionist in its approach to registration, analysis, and explanation of ecological phenomena while the other is holistic. Both traditions have considerable successes although their strengths lie in different areas. A notable example of the reductionist approach involves construction of multi-species communities from Lotka-Volterra competition and predation equations originally developed for pairs of species only. The holistic approach is usually associated with the ecosystem paradigm. It embodies a belief that a complete picture of interlinked processes and components must be obtained before sensible comments on the behavior and attributes of the system in question are possible. This belief, however well justified, is not necessarily reflected in methodologies applied by ecosystem researchers. In fact, claims are made that the focus on material and energy fluxes is quite reductionist in its essence. Without debating this issue, the discussion on paradigms highlights the different kinds of information the two approaches provide.

Scales

The scope of ecological research and its application involves an understanding of scale. For example, global ecology is concerned with interactions that take place at very large time and spatial scales among the biosphere, atmosphere, and hydrosphere, worldwide dispersal of organisms, and broad regional comparisons. By contrast, ecology of a termite digestive tract, with its community of cellulose decomposing organisms, represents the opposite size extreme. Both study objects offer legitimate research targets for ecology. The fact that both are legitimate and that both can be shown to offer valuable and important insights does not mean that all ecologists agree on their importance. A fairly vigorous debate has thrived over the last decade and still continues on the relative merits of studying large and important ecological systems as compared to small model systems, such as communities in laboratory beakers, pitcher plants, or environmental chambers. This is really a debate over whether small studies can be scaled up and large studies scaled down, which also involves the application of generalizations to specific contexts.

It appears that small natural and experimental systems offer advantages in revealing or illustrating principles of ecology while large-scale studies offer examples of applications of these principles in a context more easily assimilated by the general public. Gause's (1934) experiments showing coexistence of competitors in heterogeneous environments are elegant and simple but virtually impossible to replicate with large animals such as deer or elephants. At the same time, principles of competition and predation have been well incorporated into understanding the role of elephants in maintaining a mix of open savanna and gallery forest, the landscape feature crucial to the maintenance of megafaunal diversity in sub-Saharan Africa (Lewin 1986).

Methods

Ecologists strive to achieve a methodological rigor that matches that of physics, through published debates and informal discourse (cf. Platt 1964; Peters 1980; Fahrig 1988; Murray 2001). However, there is some doubt about the extent to which this is possible (Lawton 1999; Allen and Holling 2002). Regardless of this debate, the hypothetico-deductive method is strongly held as being at the core of current research, funding criteria, and publication streams in major journals of ecology. Experimental design, inferential statistics, and replicability of results are *sine qua non* facets of any respectable study. However, we now realize that the hypothetico-deductive method is appropriate only for some kinds of theory testing and that there are many other components of research that benefit most by using other methods.

In fact, the emphasis on "strong inference", with its criteria directly tied to hypothetico-deductive hypothesis testing, has been so muscular that it inspired a dose of corrective measures (cf. Pickett *et al.* 1994). These authors point out that a suite of methods, such as comparative, synoptic, and exploratory, in addition to the hypothetico-deductive method can contribute synergistically to the growth of ecological knowledge and that the excessive attachment to the hypothetico-deductive approach may hinder this growth.

Results

A key attribute of communicating the results of research involves the dichotomy of *uniqueness* versus *generality*. One important determinant along the gradient of uniqueness and generality is *contingency*. Individual ecological systems such as populations, communities, habitats, or landscapes are products of both general forces and contingent forces. The contingent forces include history, rare events, initial conditions, and predisposition to respond in a particular way due to, for example, the evolutionary equipment of individuals or species constituting the system. For example, how a forest responds to a windstorm may be contingent on its having been exposed to insect attack in the past. Or the response of a plant to an herbivore may depend on whether its chemical defenses have been stimulated by a prior attack or not. Given the many factors contributing to the observed pattern or affecting expected responses, one should then beware of *single cause explanations*. Such explanations may be appropriate in some circumstances but are likely to be suspect in the great majority of ecological situations, given the contingency inherent in systems. Still another attribute of results involves completed versus in-progress investigations. For example, ecosystem ecologists are quite confident that adding phosphorus to freshwater lakes causes eutrophication, barring the special circumstance where phosphorus concentration is already very high. The generalization concerning phosphorus and eutrophication was achieved by examining many cases over many years. By contrast, coral bleaching is a newly discovered phenomenon. Initial hypotheses were that climatic warming in general, and the rise of ocean surface temperatures in particular, are responsible for coral stress and that this stress results in bleaching. The case of coral bleaching provides a hint of an emerging generalization and a level of excitement and revelation as new data accumulate, but it may require major revision as we learn more about the phenomenon in question. A change of interpretation about coral bleaching due to subsequent research would not be seen by ecologists as a failure but might arouse doubts among a general audience. In fact, such a change is in the making but we will leave it to the interested readers to discover the details (see Douglas 2003).

Implications for education and public knowledge of ecology

Scientific ecological knowledge not only accumulates, but also changes due to changes in approaches, scales of investigation, methods and interpretation of results. The nature of these changes in content and the science of ecology is not entirely understood by ecologists themselves and analyses continue as to whether the change is dramatic and upsetting or gradual (Castle 2001). It is important to convey the message that most of the time, although not always, the science of ecology involves refinement, replacement of components that were faulty, and addition of new facts and concepts without, however, shaking the foundations of the continuously growing body of understanding.

For example, the original concept of community succession was strictly directional and based on competition between later arrivals and early invaders. The basic concept of community change has persisted in spite of the fact that results have shown that directionality and competition are inadequate descriptors of the process. Indeed, being able to accommodate probabilistic changes in community composition and a multitude of causes beyond competition has strengthened the theory of succession. It remains one of ecology's core concepts in spite of, or perhaps because of, the refinements to its theoretical framework. However, understanding that this change has occurred and that the conceptual constructs for community succession have changed has implications for the application of this knowledge in management and education.

Most scientists are comfortable with the notion that few things are certain and that scientific understanding is constantly evolving. The public, on the other hand, seems very uncomfortable with uncertainty and very eager to view science as unchanging fact. This gap may be impossible for ecology to tackle on its own as it may reflect a general divide between the practice of science and the public perception of scientific knowledge (Pickett 2003). Nevertheless, bridging this gap may be one of the most important tasks ahead for education. It may be much less important what the facts are, than the ability to understand how the content and methods of ecology change over time.

The scientific methods employed in ecology, together with the evolution of ideas and debates about their merits may become weaknesses insofar as their transfer to the general public is concerned. Talbot (1997) gives examples of how vulnerable ecological notions may become when they are employed out of context or with some cynicism. For example, ecologists established that some disturbance (natural!) increases biodiversity. It is easy for some to drop or omit the "natural" and paint all human disturbance as desirable. Similarly, ecosystem and community models often invoke the concept of functionally redundant species and tie this redundancy to various aspects of stability and persistence of ecological systems. Again, an unscrupulous (or naive) person might attempt to argue that redundancy should be eliminated. This advice would only pretend to be grounded in the science of ecology and, in reality, would represent a serious distortion of the message the science offers. The more complex and multifaceted ecology is, the greater the potential of its misuse through superficial sampling of its findings, particularly that non-scientists' interest in ecology is frequently linked with decision-making or action (Jenkins 2003).

CHANGING SOCIAL ASPECTS OF ECOLOGY

The public knowledge of ecology encompasses a range of misconceptions, such as believing that ecologists are tree huggers or fanatic conservationists. This results from various degrees of naivety about the scope and scientific method of ecology. The misconceptions seem to be deep and to afflict even the most educated segments of society, including academics in universities. These misconceptions may require specific targeting, depending on the audience and issues at hand and depending on whether they stem from the absence of knowledge, outdated conceptions of ecology, or biased worldviews.

Aside from the changing content and methods of ecology, there are also changing social aspects of ecology that see ecology as a collection of views held by ecologists on the state of their discipline, its needs, and roles it needs to serve. The different views on these matters by ecologists also contribute to misunderstanding of what ecology is and its role in informing management and education.

Unified versus pluralistic ecology

One prominent feature of ecology and other scientific disciplines is the relatively high degree of specialization and high degree of separation among specializations. Some see this as a virtue and attach a value-laden label of "pluralism" (e.g., Schoener 1985). Whether it is a virtue or not may be less relevant here. What is significant is that this pluralism has led to different research traditions, noted in the previous sections. While some of the plurality found in ecology may involve competing or contradictory concepts, most are compatible and subject to eventual integration. Nevertheless, the fragmentation and specialization of the discipline into many research areas pose unique difficulties for application in management and education.

This seems to be a frequent case when ecologists are asked to justify some conservation measures. For example, wolf reintroduction to the Yellowstone National Park could be argued at several levels and from several perspectives (e.g., Fritts *et al.* 1997). Depending on the individual perspective, ecologists might invoke concepts and terminology from predator–prey models, ecosystem function, or forestry management arguments in this case. Consequently, even the most common and fundamental terms may be used in ways that, while rarely contradictory, carry very different connotations. For example, the ecosystem ecologist is likely to use the term ecosystem as defined in the previous section, with focus on energy and material fluxes (Pickett and Cadenasso 2002). By comparison, a population ecologist may focus on reproductive success of a keystone species. One of the difficulties associated with a presentation of concepts from different sub-disciplines is the potential perception of contradiction or incompleteness, which becomes a source of doubts, questions, and skepticism on the part of the public.

Problems of intended and unintended pluralism continue to inspire calls and search for a degree of unification of ecological concepts. The desire to overcome fragmentation stems from practical considerations as well as from a more subtle need for conceptual generalization. Pluralism appears to be sanctioned in the theoretical realm by different idealized models that explore different aspects of the idea being considered (e.g., consider the many models on mutualism) whereas the desire to overcome pluralism at some level is reflected in a number of synthetic attempts to formulate more comprehensive conceptual or theoretical frameworks (Allen and Hoekstra 1992; Maurer 1999; Hubbell 2001) and develop frameworks that might encourage this synthesis (Pickett *et al.* 1994). Ecologists have even set up an institution, the National Center for Ecological Analysis and Synthesis, whose mission is to promote synthesis (see www.nceas.ucsb.edu/fmt/doc?/frames.html).

For educators the best recommendations we are able to offer at this stage are awareness and balance: awareness of the pluralistic and unifying views on the nature of progress in ecology is needed in presenting a broad picture of the state of the science, and balance in that both approaches are part of the way science progresses.

The nature of theory

One of the important gaps between scientists and the public involves their differing views on the notion of theory. For scientists, theory is the best conceptual framework for organizing and explaining the observable world. Ecologists have a diversity of opinions on what specifically is included in theory and how general ecological theories are developed (e.g., Fahrig 1988; Murray 1999), but they share a way of thinking about theory that may be substantially different from that held by non-scientists.

In the public sphere, theory often means something hypothetical, speculative, and subject to doubt. This is a perception gap that interferes with the effective transmission and use of ecological knowledge. The example of salmon fisheries on the west coast of North America and cod fisheries on the east coast are dramatic examples of consequences where misconceptions about science combined with flawed political mechanisms led to environmental and social disasters. In brief, ecological models that were well grounded in ecological theory led scientists to predict a marked decrease in fish populations. To the decision-makers and the general public, these were "just models," with all the "theory" underlying the models and uncertainty associated with the conclusions - a good reason in their minds to ignore ecological science.

The nature of scientific debate

To the extent that one sees scientific disagreements reported in the media, one may conclude that there is another perception gap. Scientists disagree with one another, but disagreement on a topic should not be seen as a failure of ecological knowledge. Disagreements are a normal and permanent state of affairs in science. It is the process of disagreeing that leads to various modes of testing and confirmation and, ultimately, to sound knowledge (Ziman 1978). Disagreements are more likely to be common in any new and intensely developing area of research.

Furthermore, changes in understanding resulting from disagreements are not haphazard. Usually, debates lead to new research that tests various premises and expectations and incorporates the new findings into the existing framework. For example, predator–prey cycles between lynx and snowshoe hare populations in northern Canada have a history of disagreements and subsequent refinements. The long-term time series data on predator and prey densities were gleaned from the Hudson Bay Company pelt purchase records. Early and simple interpretations were that the predator–prey population fluctuations reflected theoretically postulated dynamics (Figure 3.2). On the surface, changes in the populations had properties predicted by

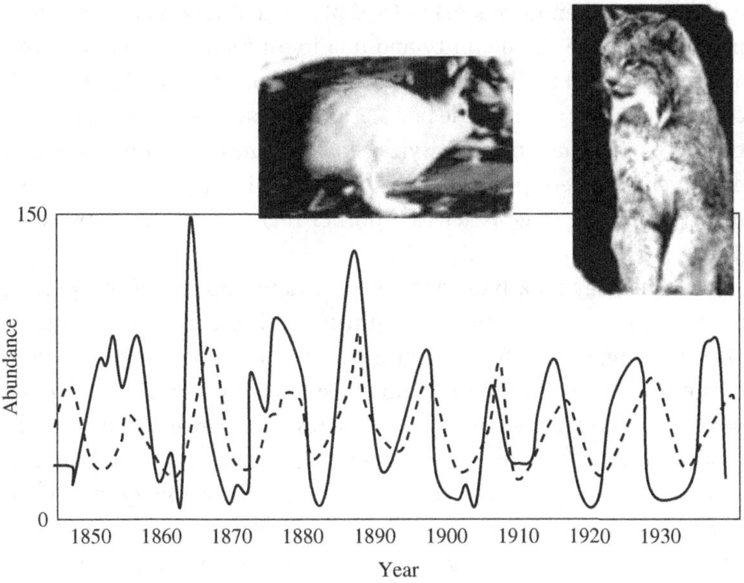

Figure 3.2. The temporal fluctuations of lynx and snowshoe hare abundance in the Canadian boreal forest. Note that the lynx population (broken line) decline follows with a small delay that of the hare populations (continuous line) as predicted by theoretical models and consistent with the idea that the success of the predator will lead to a decline of prey and a subsequent demise of the predator.

Lotka–Volterra predator–prey models. Not all ecologists agreed that the data illustrated such predator-prey dynamics and questions emerged about other possible factors that might be responsible for the observed patterns (e.g., Ruesink *et al.* 2002).

A number of alternative explanations were proposed over time. Ultimately, the causes of the observed fluctuations were elucidated. The outcome of this debate among interpretations, individual scientists, and new data has been an improved and more integrated view of the processes governing changes in the lynx and hare populations. Many of the ideas that were initially cast as doubts or alternative explanations have become recognized as contributory factors. The current interpretation of the cycles includes a number of observations and links, including: (a) as hare populations increase, grazing by hares increases (high herbivory), population of lynx begins to grow on the new food supply, high herbivory induces chemical defenses (terpene and phenolic resins) in food plants that repel the hares, both processes reduce the quantity and quality of food available to hares, and reduced food availability leads to starvation and weight loss, which contribute to the onset of the population decline; (b) large lynx population accelerates hare mortality via predation; and (c) as hare population density is reduced, predator population declines, plant populations recover, and the stage is set for another increase in hare population (Krebs *et al.* 2001).

What may look like contradictions and disarray to the general public represents a process of acquiring understanding to scientists. This meaning of debate in science needs to be exposed and communicated to the public so it is not misunderstood or abused in the political process. For example, the debates within evolutionary theory have especially been misused in the public discourse (see Eldredge 1985, particularly on creationism). Evolutionary theory has been strengthened by debates on such topics as punctuated equilibrium and species selection, rather than threatened and devalued.

The nature of uncertainty

Ecological experimentation, comparison, and inference involve probabilistic answers and uncertainty that may not be apparent in textbooks. Most scientists are comfortable with the notion that few things are certain and that scientific understanding is constantly evolving. The public, on the other hand, often seems uncomfortable with uncertainty and is eager to view scientific explanation or interpretation as concrete and definitive. However, scientific explanations and facts are often probabilistic. This implies that ecologists answer questions in terms of odds, rather than definitive statements. Practical consequences of this aspect of scientific reasoning have been particularly acute at the legal level when the public demands "yes" or "no" answers, which simply are not attainable due to the probabilistic nature of ecological phenomena. This does not mean that science cannot offer good advice, but that advice may fail to meet the legal criteria and thus end up being ineffective. Until society is well-educated enough to know that ecological evidence does not provide deterministic answers, serious problems stemming from unrealistic expectations will continue to emerge at the interface between research and public decision processes on how to use such evidence.

The phrase "scientists believe" is diagnostic of this situation. When scientists "believe" something, it usually means "scientists have established something to hold with a high degree of probability". The general public sometimes misinterprets such statements as being unsound and "hedging", and thus not worth serious consideration. In the legal arena they may be dismissed as conjectures. This gap in interpretation is undesirable and its narrowing should be a worthy goal of education.

Textbook knowledge does not help in this endeavor. Ecological textbooks emphasize conceptual and factual aspects of the discipline of ecology that are well understood, confirmed, and connected to other well-established aspects. Practicing scientists are much more aware of the evolving and probabilistic nature of this knowledge than a textbook conveys. Hence, basing ecological and environmental education on textbook knowledge may, while delivering a perfectly true picture, create a false sense of confidence that will easily break down when confronted with new situations. An ensuing crisis of credibility may have little to do with the actual scientific evidence, which may be equally good when presented in the textbook or in an environmental statement, but may result from exaggerated expectations of what research can provide.

Theory versus application

There is also a polarity between those who advocate emphasis on ecology's needs and role as a fundamental science and those who see the imperative of ecology as to help in solving environmental problems. Research priorities are likely to be affected by the outcome of the debates on the relative importance of fundamental versus applied research. Because the outcomes of the debates affect priorities, it is not surprising that ecologists involved in distinct areas of research have different stakes in promoting applied and fundamental research. This leads to the genuine competition of views often expressed in new-targeted research, funding campaigns, and special publications.

The tension between applied and fundamental research affects the gradient of generality mentioned in previous sections. Applied research is likely to lead to a focus on unique systems and more detailed investigations and thus pull ecology in one direction along the continuum of unique versus general results. Research gravitating toward fundamental issues is likely to produce results of more general nature. At the same time, applied research success may attract better funding and thus indirectly assist ecology as a whole.

Irrespective of the merits, the exposition of ecology should oscillate between applied and principle-based studies. Ecology develops best when there is a dialog between the two modes of exploration, with individual cases being powerful tests of principles forged in a more fundamental sort of research. Ecology is taught best by integrating the two in a balanced manner. What is balanced is indeed open to interpretation but the awareness of the need for balance is a good guiding rule when exposing the subject, method, and strength of ecology.

CONCLUSION

Teaching ecology requires a balanced and complete picture of the important traits of the discipline. One of these traits is richness of scope, methods, and results of ecology. This richness unfolds along a number of dimensions that include gradients of subjects, specialization, generalization, ecological systems, and spatial and temporal scales – all organized within broad paradigms. Another important trait is the scientific method that ecology implements with rigor and devotion. With the scientific rigor come the strength and limitations of scientific knowledge: its probabilistic nature, transient nature, a degree of disagreement and debate, and legacy of past paradigms.

The responsibility of ecologists is to present this diversity, development, and internal debate as completely as possible, without falling into the trap of simplistic truths that only wait to be shattered as new findings are published. The responsibility of the educators may be to convey the sense of exploration and intellectual progress that any science enjoys instead of presenting it as catalogs of results, that is, facts and principles. The two indeed need to coexist but the emphasis should be away from presenting the science of ecology solely as a collection of known facts. Bringing the struggle of ideas in its multitude of forms as an equal component of the discipline of ecology along with the factual aspects is a challenge that educators should face. This challenge will never go away because the science of ecology is governed internally by processes and criteria different from those applied by consumers of sciences in the society at large.

ACKNOWLEDGMENTS

D. Lukindo-Kolasa, a non-ecologist, suggested helpful revisions. This work has been supported by an NSERC grant (JK). Insights presented in this chapter have benefited from experience in the Baltimore Ecosystem Study, LTER funded by NSF DEB 9714835 (STAP).

REFERENCES

Allen, C. R. and Holling, C. S. (2002). Cross-scale structure and scale breaks in ecosystems and other complex systems. *Ecosystems*, **5**, 315–318.

Allen, T. F. H. and Hoekstra, T. W. (1992). *Toward a Unified Ecology*. New York, NY: Columbia University Press.

Baguette, M. and Stevens, V. M. (2003). Local populations and metapopulations are both natural and operational categories. *Oikos*, **10**, 661–663.

Begon, M., Harper, J. L., and Townsend, C. R., eds. (1990). *Ecology: Individuals, Populations, and Communities*, 2nd edn. Cambridge, MA: Blackwell Scientific Publications.

Cadenasso, M. L., Pickett, S. T. A., Weathers, K. C., and Jones, C. G. (2003). A framework for the theory of ecological boundaries. *BioScience*, **58**, 750–758.

Cale, W. G. (1988). Characterizing populations as entities in ecosystem models: problems and limitations of mass-balance modeling. *Ecological Modelling*, **42**, 89–102.

Carpenter, S. R. (1996). Microcosm experiments have limited relevance for community and ecosystem ecology. *Ecology*, **77**, 677–680.

Castle, D. (2001). A gradualist theory of discovery in ecology. *Biology and Philosophy*, **16**, 547–571.

Cuddington, K. (2001). The "balance of nature" metaphor and equilibrium in population ecology. *Biology and Philosophy*, **1**, 463–479.

Damuth, J. (1987). Interspecific allometry of population density in mammals and other animals: the independence of body mass and population energy use. *Biological Journal of the Linnean Society*, **31**, 193–246.

Douglas, A. E. (2003). Coral bleaching – how and why? *Marine Pollution Bulletin*, **46**, 385–392.

Egerton, F. N. (1973). Changing concepts of balance of nature. *Quarterly Review of Biology*, **48**, 322–350.

Eldredge, N. (1985). *Unfinished Synthesis. Biological Hierarchies and Modern Evolutionary Thought*. New York, NY: Oxford University Press.

Fahrig, L. (1988). A general model of populations in patchy habitats. *Applied Mathematics and Computation*, **27**, 53–66.

Fritts, S. H., Bangs, E. E., Fontaine, J. A., Johnson, M. R., Phillips, M. K., Koch, E. D., and Gunson, J. R. (1997). Planning and implementing a reintroduction of wolves to Yellowstone National Park and central Idaho. *Restoration Ecology*, **5**, 7–27.

Gause, G. F. (1934). *The Struggle for Existence*. Baltimore, MA: Williams and Wilkins.

Hanski, I. (1997). Metapopulation dynamics, from concepts and observations to predictive models. In *Metapopulation Biology*, ed. I. Hanski and M. E. Gilpin. San Diego, CA: Academic Press, pp. 69–91.

Hubbell, S. (2001). *The Unified Neutral Theory of Biodiversity and Biogeography*. Princeton, NJ: Princeton University Press.

Jenkins, E. W. (2003). Environmental education and the public understanding of science. *Frontiers in Ecology and the Environment*, **1**, 437–443.

Krebs, C. J., Boutin, S., and Boonstra, R., ed. (2001). *Ecosystem Dynamics of the Boreal Forest: The Kluane Project*. Oxford: Oxford University Press.

Kuhn, T. S. (1962). *The Structure of Scientific Revolutions*. Chicago, IL: University of Chicago Press.

Lawton, J. H. (1999). Are there general laws in ecology? *Oikos*, **84**, 177–192.

Leveque, C. (1995). Role and consequences of fish diversity in the functioning of African fresh-water ecosystems – a review. *Aquatic Living Resources*, **8**, 59–78.

Lewin, R. (1986). In ecology, change brings stability. *Science*, **234**, 1071–1073.

Likens, G. E. (1992). *Excellence in Ecology, 3: The Ecosystem Approach: Its Use and Abuse.* Oldendorf/Luhe, Germany: Ecology Institute.

Mahner, M. and Bunge, M. (1997). *Foundations of Biophilosophy.* Berlin: Springer.

Marks, P. L. (1974). The role of pin cherry (*Prunus pennsylvanica* L.) in the maintenance of stability in northern hardwood ecosystems. *Ecological Monographs*, **44**, 73–88.

Maurer, B. A. (1999). *Untangling Ecological Complexity.* Chicago, IL: University of Chicago Press.

Mayr, E. (1982). *The Growth of Biological Thought: Diversity, Evolution, and Inheritance.* Cambridge, MA: Harvard University Press.

McDonnell, M. J. and Pickett, S. T. A. (1993). *Humans as Components of Ecosystems: The Ecology of Subtle Human Effects and Populated Areas.* New York, NY: Springer-Verlag.

McGrady-Steed, J., Harris, P. M., and Morin, P. J. (1997). Biodiversity regulates ecosystem predictability. *Nature*, **390**, 162–165.

Murray, B. G. (2001). Are ecological and evolutionary theories scientific? *Biological Reviews*, **76**, 255–289.

O'Neill, R. V. (2001). Is it time to bury the ecosystem concept? (With full military honors, of course). *Ecology*, **82**, 3275–3284.

O'Neill, R. V., DeAngelis, D. L., Waide, J. B., and Allen, T. F. H. (1986). *A Hierarchical Concept of Ecosystems.* Princeton, NJ: Princeton University Press.

Peters, R. H. (1980). Useful concepts for predictive ecology. *Synthese*, **43**, 257–269.

Pickett, S. T. A. (2003). Why is public understanding of urban ecosystems important to science and scientists? In *Understanding Urban Ecosystems: A New Frontier for Science and Education*, ed. A. R. Berkowitz, C. H. Nilon, and K. S. Hollweg. New York, NY: Springer-Verlag, pp. 58–75.

Pickett, S. T. A. and Cadenasso, M. L. (2002). Ecosystem as a multidimensional concept: meaning, model and metaphor. *Ecosystems*, **5**, 1–10.

Pickett, S. T. A. and McDonnell, M. J. (1989). Changing perspectives in community dynamics: a theory of successional forces. *Trends in Ecology and Evolution*, **4**, 241–245.

Pickett, S. T. A., Kolasa, J., and Jones, C. G. (1994). *Ecological Understanding: The Nature of Theory and the Theory of Nature.* San Diego, CA: Academic Press.

Pickett, S. T. A., Parker, T. V., and Fielder, P. (1992). The new paradigm in ecology: implications for conservation biology above the species level. In *Conservation Biology: The Theory and Practice of Nature Conservation, Preservation, and Management*, ed. P. Fielder and S. Jain. New York, NY: Chapman and Hall, pp. 65–88.

Pickett, S. T. A., Ostfeld, R. S., Shachak, M., and Likens, G. E., eds. (1998). *The Ecological Basis of Conservation: Heterogeneity, Ecosystems and Biodiversity.* New York, NY: Chapman & Hall.

Platt, J. R. (1964). Strong inference. *Science*, **146**, 347–353.

Reiners, W. A. (1986). Complementary models for ecosystems. *American Naturalist*, **127**, 59–73.

Robertson, A. (1991). Plant–animal interactions and the structure and function of mangrove forest ecosystems. *Australian Journal of Ecology*, **16**, 433–443.

Romme, W. H. and Knight D. H. (1982). Landscape diversity: the concept applied to Yellowstone Park. *BioScience*, **32**, 664–670.

Ruesink, J. L., Hodges, K. E., and Krebs, C. J. (2002). Mass-balance analyses of boreal forest population cycles: merging demographic and ecosystem approaches. *Ecosystems*, **5**, 138–158.

Schlesinger, W. H. (1991). *Biogeochemistry: An Analysis of Global Change*. San Diego, CA: Academic Press.

Schoener, T. W. (1985). Overview: Kinds of ecological communities - ecology becomes pluralistic. In *Ecological Communities*, ed. J. M. Diamond and T. Case. New York, NY: Harper and Row, pp. 467–479.

Silvertown, J. (1982). *Introduction to Plant Population Ecology*. London: Longman.

Sterner, R. W. and Elser, J. J. (2002).*Ecological Stoichiometry: The Biology of Elements from Molecules to the Biosphere*. Princeton, NJ: Princeton University Press.

Talbot, L. E. (1997). The linkages between ecology and conservation biology. In *The Ecological Basis of Conservation: Heterogeneity, Ecosystems, and Biodiversity*, ed. S. T. A. Pickett, R. S. Ostfeld, M. Shachak, and G. E. Likens. New York, NY: Chapman & Hall, pp. 369–378.

Vitousek, P. M. (1989). Biological invasions and ecosystem properties: can species make a difference? In *Ecology of Biological Invasions of North America and Hawaii*, ed. H. A. Mooney and J. A. Drake. New York, NY: Springer, pp. 163–178.

Wu, J. and Loucks, O. L. (1995). From balance of nature to hierarchical patch dynamics: a paradigm shift in ecology. *Quarterly Review of Biology*, **70**, 439–466.

Ziman, J. M. (1978). *Reliable Knowledge: An Exploration of the Grounds for Belief in Science*. Cambridge: University Press.

DAVID R. SLINGSBY AND SUSAN BARKER

4

The role of learned societies, government agencies, NGOs, advocacy groups, media, schools, and environmental educators in shaping public understanding of ecology

INTRODUCTION

Ecology has developed from what was once regarded as "natural history" into a science with a rich array of theories, concepts, and models (McIntosh 1985). The environmental crises of the 1960s thrust ecology into the public domain, by providing evidence that environmental degradation had implications for human well-being and quality of life. Forty years later, environmental problems have developed into serious global issues. Several writers (de Groot 1992; Daily 1997; Costanza *et al.* 1997) have suggested that there has never been a more critical time for the public to understand the science of ecology, to appreciate and value services provided by the natural world, and to respond accordingly through informed decision-making.

However, there is still a gap in the public understanding of the science of ecology, despite a strong public interest in ecology and the environment, as seen by the high interest in books, films, and TV programs. There are a number of reasons why this gap exists. The academic community is perceived as being elitist, failing to communicate effectively with the public, or excluding the public by excessive use of technical language. A further problem is that the science that the public needs to know can be emerging research that is evolving. The research is likely to be considered provisional within the science community and controversial within the public domain (Gregory and Miller 1998). It is unlike the established science found in textbooks. Moreover, science in general is a long-term process, with some research taking years to draw conclusions. The time lag to provide definitive statements means that the outcomes are often no longer "news," or no longer considered to be controversial. Finally, ecology is perceived by some not to be a science at all, but an ideology (Slingsby 2001). Terms and concepts incorporating eco, ecological, and ecology, such as ecophilosophy, ecobook, ecotechnology, or ecological deficit have proliferated in recent years (Wali 1995). Few of these terms further the understanding of the science of ecology and a good number do not make ecological sense (Wali 1999). Concepts from the science of ecology have become interwoven with aesthetic, moral, or philosophical applications (McIntosh 1985; Westoby 1997).

Academic ecologists, learned societies, government agencies, non-governmental agencies, advocacy groups, media, schools, and environmental educators, have all helped shape these perceptions of ecology over time. The goal of any science education program should be to develop an understanding of the content and process of science, the nature of evidence (Tytler 2001), and to separate scientific knowledge from moral positions (Westoby 1997). In order to cope with new ideas or make decisions regarding environmental issues throughout life, a citizen also needs to understand how scientific and decision-making processes interact (Schneider 1997). This chapter will explore how these major stakeholders help shape curricula and public understanding of ecology.

SHAPING PUBLIC UNDERSTANDING OF ECOLOGY

In the past, assumptions about the public understanding of science and environment were based on a linear model of communication or information transfer. It was assumed that providing more or the right scientific information would lead to appropriate action or decision-making. Burgess *et al.* (1998) described this explanation as inadequate, since there is ample evidence that knowledge does not automatically lead to informed action or decision-making. Jensen (2002) also observed that environmental education at school traditionally focuses on knowledge transfer, rather than allowing students to actively use or internalise that knowledge. Both Jensen (2002) and Maitney (2002) argued that emotional involvement and personal experience are central to understanding and informed decision-making. Therefore, in order to increase public understanding, educators need to recognise that successful transmission or dissemination alone does not equate with successful communication or informed decision-making. The process is much more complex (Figure 4.1). A number of major stakeholders interact

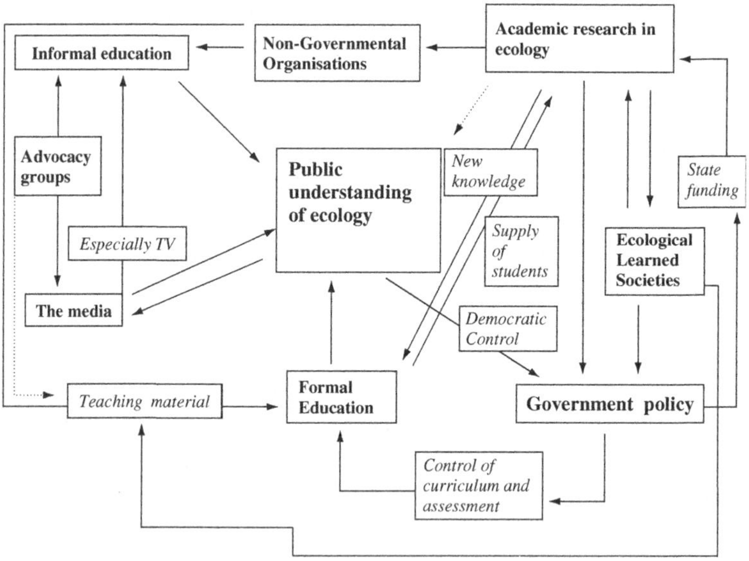

Figure 4.1. Communication model for public understanding of ecology.

and impact on the public understanding of ecology in different ways and for different purposes.

Academic ecologists and learned societies

Professional societies and ecologists are becoming more engaged in science communication and education initiatives at all levels. The reasons range from recruiting a wider diversity of students to undergraduate and graduate ecology education to maintaining public support for funding of ecological research. A commitment to attract young people to ecology requires teaching approaches and a curriculum that ensure that students have a good experience with ecology in their final years of school. This can be accomplished by ensuring that current ecological concepts and the use of fieldwork as a teaching strategy are embedded in the statutory curriculum. The British Ecological Society and the Field Studies Council have produced a report entitled "Teaching Biology Outside the Classroom: Is it Heading for Extinction?" highlighting these two critical points (Barker *et al.* 2002).

In England and Wales the final two years of school (16–18 years) are academically specialized and based on three or four subjects that the students have chosen. It is quite appropriate, therefore, to ensure that the ecological component of the biology course at this level is current, accurate, and reflects core concepts from academic ecology. However, most students do not proceed to study ecology at university. Therefore, it is also important that the public, and students, be introduced to general concepts of ecology, how ecological research is conducted, and its role in informing decision-making regarding local and global environmental issues. To be ecologically literate, from the academic or learned society perspective, one needs a framework of ecological knowledge and understanding. Ecology is understood to be about interactions and is itself synoptic, drawing on a great deal of biology together with some physics and chemistry. This knowledge of ecology should include an understanding of how ecologists work, the nature of facts, theories, and hypotheses, and the way ecological knowledge is under continual peer review. When an ecologically literate person encounters a new piece of information in a newspaper that contradicts existing thinking, they will find it interesting, evaluate the evidence, adjust their own scientific framework, and consider its implications for everyday life. An ecologically illiterate person would be thrown into confusion despite having learned a lot of scientific facts at school. Professional societies of ecology around the world have recently recognized the importance of ecological education and communication with a number of initiatives and publications. For example, the Ecological Society of Australia (http://www.ecolsoc.org.au) held a plenary symposium in 1996 at their annual meeting on communication in ecology with a resulting edited book by Wills and Hobbs (1998), *Ecology for Everyone: Communicating Ecology to Scientists, the Public and the Politicians*. The British Ecological Society (http://www.britishecologicalsociety.org) too has had a number of recent initiatives mainly for schools to stimulate interest and demonstrate the wide appeal of ecology. Careers information and a poster entitled *Ecology: A Science that Matters* show students what ecology is through what ecologists do rather than what they know. The development of a website devoted to ecological education for teachers and students and sponsorship of training in fieldwork for teachers, as well as two curriculum development projects on laboratory-based ecology (Tomkins and Dockery 2000) and on food chains (Barker and Norris 2000) have been extremely successful in bringing ecology into the classroom. The recent creation of the post of professional Education Officer is a further expression of the Society's growing commitment to engaging with young people through education. The launching of a web-based peer-reviewed compendium of more advanced ecological practical work has extended this commitment to promoting good practice in ecology teaching to the undergraduate level.

The Ecological Society of America (http://www.esa.org) also has a range of initiatives:

Teaching Issues and Experiments in Ecology (TIEE): The Teaching Issues and Experiments in Ecology Web aims to improve teaching and student learning through innovative lecture and lab activities, and to develop and disseminate peer-reviewed ecological educational curricula that undergraduate faculty can adopt and use. Two web-based volumes are available.

Strategies for Ecology Education, Development, and Sustainability (SEEDS): The Strategies for Ecology Education, Development, and Sustainability (SEEDS) Program began in 1996 as a collaborative effort to work towards increasing the number of minorities in the field of ecology. SEEDS has been shown to be an effective model for stimulating interest in pursuing ecology, providing professional development to aid science faculty in creating new ecology offerings, and taking new steps to increase cultural diversity within the Ecological Society of America.

Ecology Education Network (EcoEdNet): ESA's digital library to locate, contribute, and disseminate ecology education resources. With EcoEdNet resources teachers can incorporate images into lectures, use reviews to decide on reading materials, assign digital library articles for class discussions and journal clubs, and simulate lab experiences in preparation for hands-on live labs.

With the use of the worldwide web all these initiatives are easily accessible around the world to help promote good practice.

Government departments and agencies

Governments, particularly in a democracy, are important stakeholders in science communication and education for a number of reasons. For example, the government in the United Kingdom has a responsibility to establish and maintain national educational standards for the development of citizens with knowledge and skills to meet the future needs of society. Governments also have international treaty obligations to develop national education programs for a number of environmental protocols (e.g., Agenda 21, and the Kyoto Protocol). Therefore, government agencies play a significant role in influencing public understanding of ecology through mandated school curricula and environmental policies.

Government agencies also recognize that education can help communicate social priorities and its commitment to the environment by developing curricula that link education and government policy. For example, the Department for Education and Skills and the Department of Environment, Food and Rural Affairs in the United Kingdom, developed a curriculum on Education for Sustainable Development. The curriculum makes a very clear statement of what education for sustainable development means for the government agencies:

> Education for sustainable development enables people to develop the knowledge, values and skills to participate in decisions about the way we do things individually and collectively, both locally and globally, which will improve the quality of life now without damaging the planet for the future. (QCA 1999; Government Panel for Sustainable Development Education 1999a)

At first glance this statement sounds reasonable, but it is really a policy statement about sustainable development embedded within an education curriculum.

Governments also have an obligation to engage the public in decision-making or policy development. For example, the UK and European Community signed the UNECE Convention on Access to Information, Public Participation in Decision-Making, and Access to Justice in Environmental Matters (the Aarhus Convention; see Website: http://www.unece.org/env/pp) in 1998. The rationale behind this initiative is that leaders in these countries believe that real environmental progress can only be achieved with the participation of the citizens concerned. Ensuring public consultation will enhance public acceptance and support for the decisions to be taken. This will enable more informed and accountable decisions to be taken, and greater consensus. It will also facilitate implementation of environmental legislation. Education also has a critical role in the success of this convention.

Non-governmental organizations (NGOs)

The term "non-governmental organisations" (NGOs) no longer adequately describes the diversity of organisations outside of government that have a stake in environmental and educational policy. Rapid advances in information and communication technology (ICT) has influenced a new form of democracy and social participation. Most ICT networks are beyond institutional control and have created many new opportunities for communication that engage the public directly in debates regarding environmental issues. These new relationships and interactions, along with the traditional modes of communication by NGOs, are also creating a notion of global environmental citizenship (Barcena 1997), well beyond the traditional notions of NGOs.

Most NGOs have limited ability to influence government policy directly beyond responding to consultation documents. Therefore, public education through formal and informal education activities becomes one of the most powerful ways NGOs can influence government policy. NGOs produce some excellent teaching resources through websites, videos, computer software, and printed materials that are often free to schools. Most of the resources aim to address or enrich the requirements of the National Curriculum. However, Van Matre (2002) noted that such material could reflect bias and focus on issues in a superficial and unbalanced fashion: "Many well-meaning teachers . . ." he observed, "don't know the real origins of the material, or understand how insidious those materials are" (see Website: http://www.earthedu-cation.org/who-letter.asp). Many NGOs also rely on commercial sponsorship for funding resource development and many may have what others consider to be political agendas. Some resources are actually produced by industrial concerns such as oil companies and can easily be confused with NGO material.

The Sustainable Development Education Panel established by the Department for Environment, Food and Rural Affairs (DEFRA), a UK government department, has drawn up a voluntary code of practice for the developers of such material. Item 3 in this code requires that: "when purporting to give a balanced account of an issue, resources should accurately reflect the broad range of informed opinion on the subject" (Government Panel for Sustainable Development Education 1999b: 4). The code of practice is welcomed, but as it is voluntary the onus of identifying bias in the material still falls on the individual teacher. Teachers not only need to be trained in the evaluation of source material, but they need, in turn, to teach their pupils how to identify balanced or biased material.

NGOs also contribute to public understanding of ecology, and science in general, through recreational activities. For example, The Royal Society for the Protection of Birds (RSPB) and the Wildfowl and Wetland Trust (WWT) operate centers associated with nature reserves where the public can observe birds whilst enjoying a day out with their family. Both organizations have a strong commitment to formal and informal education, as the programs not only raise public understanding of science and appreciation of biodiversity, but also garner funding for further conservation and educational projects. The RSPB's membership is also substantial enough to give bird conservation a significant political and economic profile. Indeed one of their most popular reserves, Titchwell in Norfolk, is estimated to receive 138 000 visitors annually, contributing over £1 800 000 to the local economy (Rayment and Dickie 2001).

Advocacy groups

It is important to recognize that many advocacy groups are NGOs, but not all NGOs are advocacy groups. Advocacy groups are overtly political and often adversarial, and one should not be surprised to find unashamedly biased messages within their communication and resources. Members of advocacy groups are usually very committed to their cause and set out to arouse passion and encourage political action. Advocacy groups play a very important role in bringing issues to the forefront of public consciousness. They counteract the activities of governments attempting to cover things up, trivialize serious issues, or play down the significance of some issues. Therefore, it is important to ensure that students understand the science involved in the issues with which advocacy groups identify. Teachers and students need to be sensitive to bias and to clarify issues for themselves.

It should also be recognized that students and the public tend to evaluate scientific or ecological information in light of what the organisation represents and their motivation for disseminating information. The public often mistrusts any advice or information where there might be perceptions of ulterior motives, or if impartiality is doubted. For example, European law requires environmental impact assessments for anyone planning large developments. Experts under contract to carry out the assessments are often perceived to be in the "pay" of the developer. In these cases, students and the public need to learn how to recognize and assess the credibility of the ecological information presented.

Media

The mass media of television, radio, and newspapers exert a powerful influence on the public understanding of ecology. The media, particularly the popular press, can oversimplify complex issues and reduce them to a few words of emotive headlines. Some headlines create intense interest for a short time, leaving serious questions unanswered or giving the impression that advocacy groups cry wolf too often. The BBC, on the other hand, has a well-deserved reputation for high-quality educational programs that combine good science with excellent, inventive presentation and a commitment to public education.

However, the media does have expertise in communicating effectively with the public from which other stakeholders can learn much. It is for the other stakeholders to harness the power of the media and it is for educators to prepare citizens to cope with it.

The continuing public debate on GM technology in the UK is strongly influenced by the mass media fueled by emotive language such as Frankenstein tomatoes and suspicions of manipulation by the government and international biotech companies. GM technology good or bad is presented as a single issue rather than a series of different questions, some of which are ecological. Such a powerful cocktail of widespread public ignorance, oversimplification and paranoia sells newspapers. Advocacy groups rightly brought issues of great importance to public attention. In the Eurobarometer survey in 1996 (Commission of the European Communities 1997), one third of the UK public thought that only GM tomatoes contain genes. The poor quality of the debate was partly a reflection of the failure of the scientific community to communicate with the public. It was also due to the fact that inadequate science education left the public ill equipped to cope with the issues. The newspapers merely reflected this reality.

In the first few days of January 1993 an oil tanker named *The Braer* was wrecked off the coast of Shetland in a gale. Over 600 journalists descended on the Shetland Islands (100 miles north of the British mainland) intent on finding a story of drama, doom, and gloom. There was a wave of well-meaning sympathy for Shetland folk throughout the UK. The local newspaper, *The Shetland Times*, rightly earned an award for its coverage of a serious incident that posed an enormous threat to local wildlife and to the human population of an island community. The efforts of the rest of the media were less creditable resulting in what could be described as media pollution. The television crews stood on beaches waiting to report on a wildlife catastrophe. There were accusations that some wildlife NGOs milked public concern to raise funds and hype up the risks of oil tankers. Sadly, there was harm to local wildlife but one of the most impressive pieces of information went largely unreported. The actual damage was very much less than was predicted. The powerful wind that wrecked *The Braer* also prevented aircraft from taking off to spray detergent. The massive waves crashing on the rocks dispersed the oil slick. The wreck of *The Braer* is now itself a center of underwater biodiversity. Had the accident happened a few months later when massive populations of seabirds had returned to land to nest then there could have been a wildlife disaster of unthinkable proportion. That *The Braer* incident was a lucky escape rather than an ecological catastrophe went largely unreported. The real lesson of the accident was that there were inadequate contingency plans for dealing with a tanker in trouble and this too went largely unreported except in the pages of *The Shetland Times*. A similar oil tanker accident in south Wales a few years later suggested that, despite all the media coverage, the lesson went unfortunately unlearnt.

Schools - formal education

Each of the stakeholders noted above ultimately has some impact on school curricula and formal education. The curriculum is a statement of what society as a whole considers important for everyone's educational experience and reflects both current academic and public understanding of a subject. It also sets the standards for education and defines what is taught. Therefore, it is important that a scientific understanding of ecology is in the curriculum, but what form should it take?

In the National Curriculum of England and Wales for science, ecology is largely defined in terms of specific concepts (e.g., adaptation, food chains and cycles (QCA 1999)), whereas in the Education for Sustainable Development (ESD), ecology is defined in terms of environmental issues. The advantage of having ecology included within both statutory curricula is that the subject cannot be ignored and consequently, all pupils are exposed to at least some form of ecological education. However, in both cases the ecology is defined and assessed only in terms of its conceptual content or social context. In some cases the content can be out of date, based on ecology from the 1950s, or based on social policies that can become misconstrued as ecological concepts.

The way in which ecology is taught is dependent on how ecology is presented in the curriculum, and the individual backgrounds of teachers. There is a need to include more support for the development and assessment of practical experience or fieldwork within the formal curricula. The UK is currently experiencing a catastrophic decline in skilled personnel able to carry out ecological assessments (House of Lords 2002). There is a wealth of evidence that highlights the academic and social benefits of fieldwork (Palmer and Suggate 1996; Bogner 1998; Kollmuss and Agyeman 2002) and yet for a variety of reasons fieldwork in schools is in decline (Barker *et al.* 2002). Choi and Cho (2002) have presented evidence from South Korean middle schools that teaching biology in the context of environmental issues also enhances motivation for learning science. Developing and supporting a mandated curriculum that includes current or emerging concepts of ecology engages students in practical experiences and fieldwork, and linking ecological knowledge with environmental issues will not only nurture the development of future ecologists, but also help develop a more scientifically literate society as a whole.

A broader and more socially relevant approach to the statutory science curriculum also requires a more creative and sophisticated approach to school assessment. Current UK school examinations tend to overemphasize conceptual knowledge, although creative examination questions that also examine the application of that knowledge do exist. For example, assessing a pupil's ability to look at conflicting sources of information about an issue and to formulate an opinion already exist in History and Geography. A similar approach could be incorporated into science assessment. There is also scope to develop a means of assessing a pupil's ability to respond to an unfamiliar situation and make a decision about how they would respond, justifying themselves using scientific knowledge. Methods of assessing the effectiveness of a teaching program in terms of influencing attitudes and behaviour patterns also already exist (see Bogner 1998). To many educators the preparation of pupils for formal examinations is a barrier to creative education, yet assessment by the state in one form or another will always be with us. However, a broader and more sophisticated approach to science education might also be furthered by a more sophisticated approach to assessment.

Environmental educators

It is impossible to consider ecological education without considering how it is related to environmental education. Ecological and environmental education may appear to be almost the same thing and yet there is a chasm between them that runs deep and crackles with philosophical and political tension. Ecological education forms the scientific strand within environmental education, wherein ecology informs debate but does not claim to provide answers to aesthetic, moral, or philosophical questions. Environmental education, on the other hand, is broader and represents philosophically more diverse points of view. In some expressions of environmental education one can detect a distinctly "anti-science" prejudice that confuses science as a rational process with other forms of knowledge. Some writers are critical of the contribution that science makes to environmental education in a way that could be interpreted as reflecting an anti-science position, perhaps sometimes associated with postmodernism. Noel Gough (2002), for example, caricatures "Western" science as merely having "the appearance of universal truth and rationality." The Environmental Science Certificate of Education program of the State of Victoria (Australia), he observes, "presents an atomistic and reductionist view of large-scale ecosystem structure and function." An alternative interpretation could be that this is not so much an anti-science attitude but more a recognition of the need for a balanced and holistic approach. Robottom (1993), for example, is concerned about a dominance of "technocratic rationality" in environmental education. A technocratic curriculum with an emphasis on transmission (as propositional knowledge) of objectivist conceptions of ecology and other scientific concepts, Robottom argues, limits our understanding of environmental issues. He highlights the key problems as being the didactic transfer of knowledge as well as portrayal of a disjointed view of the environment by looking solely at the ecology in isolation from the social, political, and ethical contexts.

As Yencken *et al.* (2000) observed, science-based ecology is one of several factors which offer a global unifying framework and which enable insights into environmental matters from different cultures to be integrated. Science-based ecology education is not the same thing as education "for," "in" or "about" the environment, but provides an essential part of the environmentally literate citizen's toolbox. It is precisely because ecology is so unashamedly reductionist, atomistic, and critical when it is applied locally (i.e., in a specific situation) that it has the potential to be such a powerful unifying contributor toward thinking globally.

We are in agreement with Annette Gough (2002) when she sees the relationship between science and environmental education as a mutual one:

> Science education needs environmental education to reassert itself in the curriculum by making science seem appropriate to a wider range of

> students and more culturally and socially relevant. Environmental education needs science education to underpin the achievement of its objectives and to give it legitimate space in the curriculum to meet its goals because they are unlikely to be achieved from the margins.
> (A. Gough 2002: 1210–1211)

CONCLUSION

All forms of communication have multiple purposes (Keith 1997). Although ecological knowledge is largely derived from the activities of academics and learned societies, the public understanding of ecology is shaped by the complex interaction and feedback between the major stakeholders or users of ecological knowledge noted above. The statutory curriculum as mandated by government reflects the influence and interaction of these stakeholders and their perceptions of ecology. In the end, it is the complex interaction of these stakeholders that determines what actually is taught in the classroom.

Ecology *informs* behavior but does not judge its success by its ability to modify behavior. In their very useful review of environmental psychology, Kollmuss and Agyeman (2002) examined a question that concerns many environmental educators – namely "what motivates people to act environmentally and what are the barriers to pro-environmental behavior?" There is abundant evidence that increasing a person's knowledge of the environment does not necessarily make their behavior more "pro-environmental." Diekmann and Preisendoerfer (1992) proposed a "low cost/high cost" model – an environmentally aware person might well recycle waste but continue to use a car as an alternative to walking. Such assumptions about what is or is not "pro-environmental behavior" are not a part of what we understand by ecological education. It is a valid point of view that if everyone was careful to switch off a light on leaving a room then it would conserve energy and might encourage others to do the same. It could also be a valid point of view that avoiding using one's car is not going to make any difference to climate change unless everyone else does the same. Educating young people that certain behaviors are, from an environmental point of view, "good" and others are "bad" can be a very simplistic approach to teaching environmental ethics and has little to do with the science of ecology.

On the other hand, *understanding the ecological reasons why* such behaviors have been suggested may sensitise and equip a person with the skills to make up his or her own mind and decide how to respond to new ideas in future. Ecological education *is* about ecology as an objective value-free science but it is important that it be presented to the citizens of tomorrow in an ethical *context*. It is important to differentiate the ecology, which informs debate from what people say *about* the associated ethical issues. Diekmann and Franzen (1999) suggested that environmentally aware people who could be reluctant to make lifestyle sacrifices might, nevertheless, be predisposed to accept *political changes* such as higher fuel taxes. We suggest that this is a desirable way in which ecological understanding may influence behavior. Another desirable outcome could be said to arise when the public demands environmental leadership from their politicians. Thus a key way forward for sustainable ecological education is to place the scientific knowledge in an ethical context. The concept of sustainability is not science but an ethical precept (Norton 1992; Asheim *et al.* 2001) and stated desires for simultaneous equity, prosperity, and environmental protection represent moral positions (Shields *et al.* 2002). This in turn leads to value judgments that are crucial in guiding public policies and actions.

This chapter has emphasized the importance of ensuring that ecological education is firmly rooted in science and the science curriculum whilst also stressing the key role ecological thinking has in achieving a sustainable environment. The science of ecology contributes to Education for Sustainable Development by providing *information* needed to manage resources sustainably, communicates in a *politically neutral* manner, and *educates policy-makers* and the public (see NCSE 2000) on how to evaluate the credibility of ecological information being communicated by a wide range of stakeholders.

REFERENCES

Asheim, G., Buchholz, W., and Tungodden, B. (2001). Justifying sustainability. *Journal of Environmental Economics and Management*, **41**, 252–268.

Barcena, A. (1997). *Global Environmental Citizenship: Our Planet 8.5*. Kenya: United Nations Environment Program (UNEP).

Barker, S. and Norris, C. (2000). *Feeding Relationships: An Ecological Approach to Teaching of Food Chains in the Primary School*. London: British Ecological Society.

Barker, S., Slingsby, D., and Tilling, S. (2002). *Teaching Biology Outside the Classroom. Is it Heading for Extinction? A Report on Outdoor Biology Teaching in the 14–19 Curriculum*. Shrewsbury, UK: Field Studies Council.

Bogner, F. X. (1998). The influence of short-term outdoor ecology education on long-term variables of environmental perception. *Journal of Environmental Education*, **29**, 17–29.

Burgess, J., Harrison, C., and Filius P. (1998). Environmental communication and the cultural politics of environmental citizenship. *Environment and Planning*, **30**, 1445–1460.

Choi, K. and Cho, H. (2002). Effects of teaching ethical issues on Korean school students' attitudes towards science. *Journal of Biological Education*, **37**, 26–30.

Commission of the European Communities. (1997). *European Opinions on Modern Biotechnology, Eurobarometer 46.1.* Brussels: Commission of the European Communities.

Costanza, R. *et al.* (1997). The value of the world's ecosystems services and natural capital. *Nature*, **387**, 253–260.

Daily, G. C., ed. (1997). *Nature's Services*. Washington, DC: Island Press.

de Groot, R. S. (1992). *Functions of Nature: Evaluation of Nature in Environmental Planning, Management and Decision-making.* Groningen, The Netherlands: Wolters Noordhoff BV.

Department of Environment Food Rural Affairs. (2002). *Survey of Public Attitudes to Quality of Life and to the Environment: 2001*. London: HMSO.

Diekmann, A. and Franzen, A. (1999). The wealth of nation and environmental concern. *Environment and Behaviour*, **31**, 540–549.

Diekmann, A. and Preisendoerfer, P. (1992). Persoenliches Umwelverhalten: Die Diskrepanz zwischen Anspruch und Wirklichkeit. *Koelner Zeitschrift für Soziologie und Sozialpsychologie*, **44**, 226–251.

Gough, A. (2002). Mutualism: a different agenda for environment and science education. *International Journal of Science Education*, **24**, 1201–1215.

Gough, N. (2002). Thinking/acting locally/globally: Western science and environmental education in a global knowledge economy. *International Journal of Science Education*, **24**, 1217–1237.

Government Panel for Sustainable Development Education. (1999a). *A Strategy for Sustainable Development for the United Kingdom*. London: HMSO.

(1999b). *Supporting Sustainable Development through Educational Resources: A Voluntary Code of Practice*. London: HMSO.

Gregory, J. and Miller, S. (1998). *Science in Public: Communication Culture and Credibility*. New York, NY: Plenum Press.

House of Lords. (2002). *What on Earth? The Threat to Science Underpinning Conservation.* Third Report, Select Committee of Science and Technology. London: HMSO.

Jensen, B. B. (2002). Knowledge, action and pro-environmental behaviour. *Environmental Education Research*, **8**, 25–334.

Keith, W. (1997). Science communication: beyond form and content in scientific and technical communication. In *Scientific and Technical Communication*, ed. J. H. Collier and D. M. Toomey. London: Sage Publications, pp. 299–326.

Kollmuss, A. and Agyeman, J. (2002). Mind the gap: why do people act environmentally and what are the barriers to pro-environmental behavior? *Environmental Education Research*, **8**, 239–259.

Maitney, P. T. (2002). Mind in the Gap: summary of research exploring "inner" influences on pro-sustainability learning and behaviour. *Environmental Education Research*, **8**, 1469–1587.

McIntosh, R. P. (1985). *The Background of Ecology: Concept and Theory*. Cambridge: Cambridge University Press.

National Council for Science and the Environment (NCSE). (2000). *Sustainable Resource Management*. [Online] URL: http://www.cnie.org/2000conference/20.cfm (retrieved May 2003).

Norton, B. (1992). Sustainability, human welfare, and ecosystem health. *Environmental Values*, **1**, 97–111.

Palmer, J. and Suggate, J. (1996). Influences and experiences affecting the pro-environmental behaviour of educators. *Environmental Education Research*, **2**, 109–121.

Qualifications and Curriculum Authority (QCA). (1999). *The National Curriculum*. London: HMSO.

Rayment, M. and Dickie, L. (2001). *Conservation Works for Local Economies in the UK*. Sandy, Beds, UK: Royal Society for the Protection of Birds.

Robottom, I. (1993). The role of ecology in education: an Australian perspective. In *Ecology in Education*, ed. M. Hale. Cambridge: Cambridge University Press, pp. 1–9.

Schneider, S. H. (1997). Defining and teaching environmental literacy. *Trends in Ecology and Evolution*, **12**, 457.

Shields, D. J., Solar, S. V., and Martin, W. E. (2002). The role of values and objectives in communicating indicators of sustainability. *Ecological Indicators*, **2**, 149–160.

Slingsby, D. R. (2001). Perceptions of ecology: bridging the gap between academia and public through education and communication. *Bulletin of the Ecological Society of America*, **82**(2), 142–148.

Tomkins, S. and Dockery, M. (2000). *Brine Shrimp Ecology*. London: British Ecological Society.

Tytler, R. (2001). Dimensions of evidence, the public understanding of science and science education. *International Journal of Science Education*, **23**(8), 815–832.

Van Matre, S. (2002). *Letter of Introduction to Earth Education*. [Online] URL: http://www.eartheducation.org/who-letter.asp (retrieved May 2003).

Wali, M. K. (1995). Ecovocabulary: a glossary of our times. *Bulletin of the Ecological Society of America*, **75**, 106–111.

(1999). Ecology today: beyond the bounds of science. *Nature and Resources*, **35**, 38–50.

Westoby, M. (1997). What does "ecology" mean? *Trends in Ecology and Evolution*, **12**, 166.

Wills, R. and Hobbs, R. J. (1998). *Ecology for Everyone: Communicating Ecology to Scientists, the Public and the Politicians*. Sydney, Australia: Surrey Beatty and Sons.

Yencken, D., Firn, J., and Sykes, H., eds. (2000). *Environmental, Education and Society in the Asia-Pacific: Local Tradition and Global Discourses*. London: Routledge.

Part II Changing perspectives of education

In this section Bob Jickling shares a personal story as a schoolteacher and researcher on the difficulty of differentiating between education and advocacy in environmental education. Jickling provides examples of educational approaches that can be perceived as advocacy and offers "guideposts" for educators on how to deal with controversy inherent in environmental education practice. Joy Palmer and Joanna Birch provide the reader with an overview of changing perspectives and approaches in environmental education research. These authors highlight some of the gaps between the environmental research perspectives and environmental education in practice. They conclude with examples from current research that are closing these gaps. The section concludes with a historical review by John Disinger on the changing purposes of environmental education. Disinger reminds us that concerns with advocacy are not new and highlights how various groups have influenced development of environmental education and the underlying education–advocacy debate over time. He concludes with how educators and schools should move forward in developing environmental literacy.

BOB JICKLING

5

Education and advocacy: a troubling relationship

INTRODUCTION

Environmental education and environmental advocacy are a troubling pair. Their relationship has been uneasy for many, including me. How does a person work on behalf of what she or he cares about - but in an educational way? Can you? If you remove care from the equation can you really have an educational experience? Or, if you want people to care - about each other, the environment, ideas, and noble action - can education play a legitimate role? These are questions for teachers, whether explicit or implicit, in the hundred little actions that occur each day in a classroom. And, they are just as important for those working in other educational settings with interpretation programs, non-governmental organizations (NGOs), and government agencies. Questions like these have troubled me for a long time and have been fueled by a number of experiences. Two stand out as particularly important, the first as a schoolteacher and the second as a researcher.

A teacher story

As a Whitehorse schoolteacher in the early 1980s, I had to examine my role in the midst of a controversy about wolves. The Yukon government had initiated a wolf kill program. I wanted my students to get involved in the issue, and participate in the debate. However passionate my feelings, I was deeply troubled by a lack of philosophical guidance and curricular options; I had received little preparation for such a task. I knew also that every day I faced a class with individuals of different cultural backgrounds and values, some of whom had parents who supported the activities about which I was so skeptical. My instincts told me that imposing my views, or proselytizing, would be neither practically viable nor educationally justifiable.

I looked for inspiration in available environmental education materials and found none satisfactory. While many of these, including a locally developed package, spoke eloquently about addressing "value" questions, none presented activities with any real promise of achieving this goal. I was disappointed. I was also dissatisfied with my lack of understanding about what environmental education was, and what it could be. Many years later, I see that environmental education is evolving, broadening in scope, and penetrating ever more deeply into the tangled underbrush of values, politics, and ethics. But, it seems there is still much more to do.

So far I have given a glimpse of my story, but mine is not the only one. Systematic removal of wolves is not new in the Yukon; it is part of our history. The first biologist hired for the Yukon in the 1950s was given the task of overseeing a widespread wolf-poisoning program (Gilbert 1994). The scale and methods of this program would be unthinkable by today's standards. Nevertheless, it is far from clear that the new science of wildlife management rests on assumptions that are fundamentally different from those of Yukon's pioneer biologists. Marching in accordance with Bacon's dictum, "the secrets of nature reveal themselves more readily under the vexations of art than when they go their own way" (Bacon 1620, in Anderson 1960: 95), wolf controls are now dressed with the "respectability" of science, the "objectivity" of experimental design, and vexatious treatment conditions. Unfortunately, this presumed respectability masks important philosophical underpinnings.

There are also "frontier" stories such as the one provided by the Yukon Fish and Game Association and its members. This is a Euro-Canadian hunting tale involving, in varying proportions, the pursuit of food, pleasure, and trophies. Here the wolf is a competitor. Last, yet first, there are the voices of Yukon First Nations whose people have inhabited this place, now known as "Yukon," for a long time. Theirs is a story of life inextricably linked to the land – their home with wolves and caribou as part of this land.

The stories are interesting in their own right and the storytellers are, whether stated or not, shaped by different ideas and different values. But which story should we follow? And what role can education play? These were the kinds of questions I asked 20 years ago, and they seem just as relevant today.

A researcher story

The teacher story and researcher story are connected. The educational dissonance experienced in the face of a local controversy, the lack of available guidance, and some intellectual curiosity about the whole tangled issue were important motivations for enrolling in a Ph.D. program.

A few years later, while conducting my dissertation research, I interviewed a subject who presented me with more to think about – another sharp twist in a set of already difficult questions. When asked about the role of interest groups in educational settings, he said:

> How would Greenpeace, for example, fit in with environmental ed. in the school, or Friends of Wolves, or other fanatical interest groups . . . if a person like that was given a contract to talk to Yukon students about environmental ed., I would be very upset. I don't see how it can fit in. (Jickling 1991: 176)

What makes this story interesting is that this person wasn't a resource developer, or anti-environmentalist. He would, in fact, describe himself as a conservationist. He had an active interest in natural history, the protection of wildlife through enforcement of existing legislation, and ensuring a good, conservation-minded hunting and fishing ethic. For this person, environmental education should include more natural history. And this might be supported through the careful preparation of wildlife specimens and construction of natural history displays. Throughout our interview he portrayed consistent respect for the land.

At about the same time, I was making a philosophical exploration of the concept of education – its nature and role in preparing thoughtful and critical citizens. Here I found educational philosopher R. S. Peters (1973: 256) suggesting that it would be unreasonable "to deprive anyone of access in an arbitrary way to forms of understanding which might throw light on alternatives open to him." This comment, juxtaposed against those of the research subject, framed another important educational question. How can we ensure that educational programs provide a sufficient breadth of alternatives for learners to ponder, and use to construct meaning in the face of important decisions, to ensure that they are adequately prepared to play an active role in democratic processes?

While the research subject in this story was a thoughtful citizen who was concerned about his environment, he had his limits. But what if environmental thinking needs to transcend the boundaries of conventional thinking and encounter more thought provoking, even radical, ideas? How do we enable our students to push beyond the bounds of our own best thinking, or the conventional wisdom of the day? How do we ensure that they can be exposed to more alternatives? I'm certainly not suggesting that Greenpeace is of preeminent importance. But I am saying that we may need to open the door a little wider in order to enable thoughtful consideration of more "radical" ideas that take us closer to "root" causes and concerns.

The larger goal of this chapter, then, is to explore the idea of "caring," and the kind of advocacy that grows from it, and their relationships with education. Twenty years later, my original instincts, to avoid proselytizing in the name of education, still seem like good ones. I do hope, however, that the chapter will provide a few additional guideposts for those who share, or are about to share, similar stories.

SO WHAT'S NEW?

A few things have changed over the past couple of decades. Whereas I could find only a handful of curriculum supplements that were concerned with environmental issues in the early 1980s, there is no shortage now. An environmental educator today can easily find curricula and learning packages on protected areas, biodiversity, wolves, wild sheep, endangered species, climate change, wilderness, urban environments, water, forestry, rivers, mountains, oceans, contaminants, and more. For good measure, curricula on forestry, mining, and oil production are also available. And then there are curricula about human rights, development, globalization and so on. This is just the tip of an iceberg.

A friend working with the local Department of Education reports being literally inundated with dozens of new curriculum initiatives each year. So we have learned a lot about the theory and practice of curriculum development. Some (e.g., Pratt 1980) have produced formulas to work from and we now do needs assessments, organize writing workshops, conduct field trials, revise and refine new curricula. Some initiatives have been considered worthy of implementation. Sponsors of Project WILD, for example, require educators to take a workshop on the use of their learning materials before being provided with a free copy of this learning supplement. Others just provide boxes of curricula for distribution to school principals, libraries, contact persons, or anyone else who will listen.

Some time ago, E. F. Schumacher (1973) in *Small is Beautiful* raised doubts about the efficacy of Western education. Despite the widespread belief that education has a key role to play in the resolution of our problems he wondered if the quality of education keeps pace with the quantity of educational materials generated.

In our contemporary context Schumacher's observation seems to be being played out in a kind of curriculum "tug of war" where competing interests struggle to woo educators and to engage impressionable young minds with their own "messages." No doubt, the possibilities for assistance are much greater than 20 years ago when I struggled as a classroom teacher. And many educators and interpreters have been well served by these efforts. Still, I'm left with the questions: Is this the way to go? Do we really need more curriculum packages, prepared according to conventional understanding about curriculum development? Does this vast array of material lead us to inquire deeply about the questions that most motivate us?

In addition to myriad sets of learning materials we have had a parallel development of "new educations." In earlier times we had nature education. Since then there have been permutations and combinations of (in no particular order) outdoor education, conservation education, environmental education, experiential education, humane education, development education, global education, peace education, human rights education, and various manifestations of religious education. Now we have a newer breed, the "for" educations. These include, for example, education for sustainable development, education for sustainability, education for Deep Ecology, and education for red/green environmentalism. Recently, after years of working to reorient environmental education toward education for sustainable development and/or education for sustainability, I learn that in the minds of some these are no longer the same but rather two distinct entities (McKeown and Hopkins 2003).

What do these various, and sometimes contesting, educations have in common? What motivates them? Are we just inventing more educations but not really achieving meaningful educational change? Are we now producing piles of curriculum supplements and educational variants but with diminishing returns – educationally, and in terms of energy and the limited resources consumed?

One thing that does unite the multiple educations is that they are all slogans, and this is not necessarily pejorative. According to Scheffler (1960), educational slogans are rousing and can provide rallying symbols for key ideas, attitudes of educational movements, and overlooked priorities. They also express and foster community of spirit, attracting new adherents and nurturing veterans. While slogans are largely symbolic and representative of ideas and attitudes that may be expressed more fully elsewhere, Scheffler suggests that they can take on a life of their own. I think that most readers can understand the rallying and nurturing aspects of these slogans. But can we continue to be pulled in ever more directions by ever more educations? And what are the ideas and attitudes that these slogans represent? Are there commonalities between these educations? What happens when slogans take on a life of their own, slipping toward more advocatory positions? These are the questions, beginning with the last, that I would like to explore in the next sections.

WHEN SLOGANS GO WRONG

In my own first attempts to understand differences between education and environmental advocacy, I sought to distinguish between education on one hand and training and more doctrinaire activities on the other (Jickling 1991, 1992). Distinctions can identify a range of possibilities and illustrate difficulties. In this sense, some of the more blatant cases of misplaced advocacy can be identified. Like other concepts, advocacy does not exist as a clear concept with a fixed meaning. Rather, it occurs in greater or lesser degrees. However, strong advocacy is often associated with increasingly forceful pleas for a case, direction, or ideologically grounded position.

Much early work in environmental education was couched in terms of observable problems such as environmental planning, pesticide use, community blight, air and water pollution, traffic congestion and so on (Stapp *et al.* 1969). Eliciting action on these and other issues seemed to define much of the agenda. Aims were often couched in terms of producing informed and skilled citizens willing and able to take action to resolve environmental issues, or promoting the acquisition of responsible environmental behaviour (e.g., Hungerford *et al.* 1980; Sia *et al.* 1985–86).

Common responses to this agenda suggest instrumental and behaviorist predilections. That is, education can be used as a means to an end, and behavior modifications could be an appropriate tool. For example:

> While the pathway represented ... by knowledge, skills, and personality factors is the more desirable pathway to encourage environmentally

> responsible behavior, it may be more efficacious, in the case of certain environmental problems, to manipulate situational factors in order to produce desired behavioral changes. (Hines *et al.* 1986–87: 8)

It wasn't always clear how responsible behaviors were determined though, in some cases, they were defined by reference to Sierra Club membership or similar activist populations (Sia *et al.* 1985–86).

The problem for many educational researchers is that it is just not clear how such talk of "producing desired behavioral change" to achieve a "desired state" can be considered educational. Such talk is more suggestive of training, where skills are perfected through repetition and practice, and minimally involved with understanding; the aims are instrumental and intent on producing a desired end.[1]

Another distinction between education and more prescriptive or advocacy-oriented activities is manifest in the term "education for sustainable development" (Jickling 1992; Jickling and Spork 1998). The report of the World Commission on Environment and Development (1987) in *Our Common Future* first links sustainable development with education. Perhaps the clearest indication of a strong advocacy orientation of "education for sustainable development" can be seen in *Our Common Future*:

> Sustainable development has been described here in general terms. How are individuals in the real world to be persuaded or made to act in the common interest? The answer lies partly in education, institutional development, and law enforcement. (WCED: 46)

This statement suggests that sustainable development is in the common interest and that the public must be persuaded or made to pursue this end. Further, education can contribute to the process of persuasion required.

Despite years of critique, adherents of sustainable development continue to advocate this agenda as an educational end. Education is still seen as a tool in the "critical endeavor of attaining a sustainable future" and that it "should be able to cope with determining and implanting these broad guiding principles [of sustainability] at the heart of ESD [education for sustainable development]" (Hopkins 1998).

The principal problem with these examples is that education is presented as an instrumentalist and ideological tool; it is somehow educationally justifiable to implant learners with the guiding principles of sustainability in the service of sustainable development. Highlighted in this way, many educators find these sentiments a misrepresentation of their task. They see their job as enlarging students' ability to think freely and creatively, rather than prescribing what learners should think. Imagine implanting the principles of Marxism into the heart of an American educational system. This would be seen as indoctrination. The same should also hold true of less overtly contentious ideologies such as sustainable development. Again there is a type of advocacy promoted which seems at odds with ideas about education. Though, as we shall see, the relationship between these concepts is more complex than the examples suggest.

Another example of a strong advocacy orientation extant in the literature can be found in John Fien's (1993) *Education for the Environment*. In this text he discussed a variety of philosophical and political visions of the future based on differing approaches to environmental issues. He then decided, based on his analysis of these differing approaches, that a particular variety, which he characterized as a "red-green" (or ecosocialism) vision, showed the most promise. He then claimed that the desired "red-green" future lay at the heart of education for the environment. While Fien is not alone in formulating a "correct" or "recommended" vision and politics at the core of environmental education, his work typifies a trend by some toward an assertive commingling of personal commitment and the role of education.

It is one thing for an individual to assess the range of available environmental alternatives and then advance a case for a preferred option; it is quite another to insert this option into the heart of anything educational. The first function is socially vital. There must be people who think hard about the future, challenge the conceptions of others, and then advance ideas of their own. These are the thoughts and processes that provide our society with the content and motivation with which to strive for a better future. However, it is not at all clear that the second function is educational. It is not clear that we should privilege our preferred option, or agenda, as the aim and that it is appropriate to advance this "agenda" as the aim of, or heart of, or as central to, education.

If the approaches presented above are at odds with ideas about education, then we will certainly have critics. Some will deny or downplay environmental issues and vigorously denounce the inculcation of environmentalism into the minds of young citizens (Sanera and Shaw 1996; Sanera 1998). In the context of my own story, there are problems of another kind. How can an educational environment be created where students in the Yukon (and elsewhere) can be introduced to ideas outside of the mainstream political spectrum? How can they be given the conceptual and practical tools that could enable them to move beyond the standards set by world leaders (e.g., WCED 1987; UNCED 1992; Scoullos 1988)? How can they have the opportunity to develop innovative and new ideas?

Unfortunately, inserting the sustainable development, sustainability, red-green, and other advocacy-oriented agendas into environmental education is problematic. Educators are concerned about demands for obedience and acquiescence in the face of hegemonic discourses. As Canadian philosopher John Ralston Saul (1995: 190) says:

> Equilibrium, in the Western Experience, is dependent not just on criticism, but on non-conformism in the public place.

Here equilibrium represents that civic conversation that occurs in fully functioning civil societies and that prevent them from sliding toward ideology, toward a "you're with us or you're against us" society. According to Saul, criticism is important. But, so too is non-conformism. It is important to make a little space for those so-called "radicals," and other creative perspectives in the face of enormous pressure to conform within a narrow band of social standards. When slogans go wrong, there is a tendency to drift toward less, rather than more, alternatives and toward more ideologically loaded and advocacy-oriented positions. There is little tolerance of dissent, and there is even less room for non-conformism.

It is against these concerns that it first seemed important for me to distinguish between education and advocacy. Not only did I want to be able to answer critics, but I also was concerned that environmental education should be able to develop as environmental thinking itself continues to develop (see Weston 1992). However, distinctions are simply devices that we can employ to see an issue from alternative perspectives; they aren't truths. The examples given illustrate some problems with advocacy-oriented positions. They begin to establish territory for discussions. However, there are nuances to explore in complex and real circumstances.

MODERNIZING ENVIRONMENTAL EDUCATION?

The title of this section was taken from a presentation made by Walter Leal Filho (2003) at the First World Environmental Education Congress. Though the title held much promise, the analysis offered little new; we should perhaps move a little more in the direction of sustainability, but essentially environmental educators should just keep going in much the same directions as they have been.

But what if we were to take seriously the idea of modernizing or, even better, reimagining environmental education? What would that look like, if we were to dig a little deeper? What if we were to depart from past trends that have led to devising evermore adjective driven educations? Or, what if we stopped looking for replacements for "sustainable development," "sustainability," "red-green environmentalism," and other aims *for* education? What then?

Consider why we have created so many different forms of education: nature education, outdoor education, environmental education, experiential education, and global education along with so many others? What educational priorities are these slogans directing us to think about? Are there some commonalities amongst them?

Returning to E. F. Schumacher (1977: 1) again, this time from *A Guide for the Perplexed*, we might find that he too had been asking similar questions when he said:

> All through school and university I had been given maps of life and knowledge on which there was hardly a trace of many of the things that I most cared about and that seemed to me to be of the greatest possible importance to the conduct of my life. I remembered that for many years my perplexity had been complete; and no interpreter had come along to help me. It remained complete until I ceased to suspect the sanity of my perceptions and began, instead, to suspect the soundness of the maps.

I suspect these words resonate with many readers. Framed this way, we might think about reimagining environmental education – and many of the other educations – as a journey into blank spots on the curricular maps of schools, university, and life. First, knowledge: what ways of knowing are missing? What do we care about that is absent from schools, universities, and that ongoing curriculum of lifelong learning? Second, there seems a wide gap between much curriculum material and life itself, especially in contemplating how to conduct a good life and how to live well.

If, following Schumacher, we shift our emphasis to thinking about epistemological blank spaces and how to live well in a place, then we can begin to look at educational trends anew. When we do, we find one thing that the various approaches discussed here, and their emergent slogans, share. They are all protesting against the status quo; they are all protests on behalf of values.

Talk about the status quo is ubiquitous; we all complain about the status quo, but what force drives it? What authority serves to maintain this corpus of social conventions and values? Canadian environmental philosopher Neil Evernden (1985) provides an important clue. Drawing on work from others before him, he suggests that *the real authorities in a culture are unquestioned assumptions*. Our guiding and stabilizing authorities are the values and beliefs that are buried so deeply in our belief systems and lifestyles as to be transparent - unrecognizable on a daily basis. Any real change will require disrupting these assumptions. What we really seem to want - or need - is to put the "status quo" up for debate.

This is a theme that was on the minds of the 5500 Canadians who participated in Environment Canada consultations between 1999 and 2002. In fact, they felt strongly that environmental learning must be inextricably linked to values and ethical ways of thinking (Government of Canada 2002). Canadian environmental educators have said, in essence, that learning opportunities should include examination of the belief systems that guide relationships between people, societies, and the more-than-human world. This means examining cultural assumptions and what for many is the process of ethics, or ethical inquiry.

Of course, this is easier said than done. But for starters let's go back to John Ralston Saul (1995) who reminds us that the vibrant civic conversation that occurs in fully functioning civil society depends on criticism and non-conformism. In the absence of these qualities there is nothing to prevent us from drifting toward social conformity and uncritical acceptance of the status quo and the ideologies upon which it rests.

In putting these ideas about criticism and non-conformism into an educational context, consider the sentiments of Richard Shaull (in the Foreword to Freire 1970: 15) and many others:

> Education either functions as an instrument which is used to facilitate the integration of the younger generation into the logic of the present system and bring about conformity to it, or it becomes "the practice of freedom" the means by which men and women deal critically and creatively with their reality and discover how to participate in the transformation of their world.

In this sense, education is not about social reproduction, but rather it is about creating the ability to engage in a serious conversation about social values and the status quo, and ultimately to transcend social norms. When thinking about educational programs, and even what seems to be the best that environmental education has to offer, we can ask, does the material disrupt the status quo or reinforce it?

Now we are getting somewhere, educationally speaking. We seem to be locating some of those blank spots on Shumacher's curricular maps. This is almost certainly the same space that so many others have been seeking in their journeys to highlight overlooked educational priorities. This isn't just a rhetorical argument. The vast majority of the 5500 Canadians consulted stated that environmental education needed most to be concerned with values and ways to bring them into our ethical and political conversations.

In terms of my own story, as a public school teacher concerned about wolf issues, the results of these Canadian consultations are affirming. I seem to be in good company. There are, indeed, poorly charted portions of our curricular maps, vast spaces with just a few trails leading into the terrain of values, ethics, and politics.

However, as important as my public school experience was, my interest in wolves can probably be traced to a ski trip almost a decade earlier. Crossing a frozen lake one cold winter evening, I encountered a wolf pack. I didn't see them in spite of the full moon in a starry sky. I remember how the night's silence was pierced by the ear-splitting cracks as the lake's surface adjusted to the plunging temperature. And then there were howls from the wolves, howls that sang from all around, then reverberated from the densely forested hillsides. I cannot fully describe or interpret the experience, shared by two weary skiers, but I remember it like yesterday.

There have been many more lupine experiences in wild places of the Yukon Territory: the mother watching me curiously from the banks of the South Macmillan River while her two cubs skipped about playfully unaware of this human presence, the unmistakable odor of wolf as I wriggled into a recently vacated den dug 5 meters into a sandy bank, or tracks in the sand outside my tent that were not there the evening before as I climbed inside for the deep slumber of the weary traveler.

For me, wolves have been a reality for a long time. They have become part of me through these experiences. Caring about wolves, rooted long ago, remains part of my personal and political activities. Yet, while wolves motivate these activities, they are about more than that. Wolves are both real and metaphoric. They are subjects of care, and they are symbols of a struggle to understand our human place in a more-than-human world. And what place is there for care in education? Is this another undermapped spot on curricular maps?

When we talk about values, ethics, and politics we cannot be simply objective; there is always a more subjective and emotional component. For those of you who recoil at this thought, I suggest that the very act of recoiling is an emotional response. To attend to these emotional dimensions of our lives is to be human. Think for a moment about what leads people to careers in environmental or social activism, or even science. Is it a joy found in theoretical modeling of population dynamics? Or is it a natural empathy toward natural and human worlds, a love of natural history, or passion and care for people and places? Is it perhaps the howl of a wolf on a moonlit night, or myriad other intimate experiences that define who we are?

Yet, do we make room in our curricula for emotional understanding, for nurturing care and feelings? Does emotional understanding figure in standardized testing? Can it? Is there sufficient breadth in the knowledge systems that underpin our curricula? The venerable Norwegian philosopher Arne Naess (2002: 51) pointed toward an answer to some of these questions when he said:

> The rationality that characterizes the knowledge society is of an extremely limited kind – a petty rationality – that does not ask what are our most fundamental priorities and values as human beings. This rationality has lost sight of our aims and is merely concerned with means. Is the conflict between emotion and rationality real? And is it not often imagination and emotion that drives scientists and scholars?

It is this petty rationality that we are left with when we allow educational programming to skirt questions of values, ethics, politics, and feelings. Put another way, Aldo Leopold (1966: 261) said:

> It is inconceivable to me that an ethical relation to the land can exist without love, respect, and admiration for land and a high regard for its value. By value, I of course mean something far broader than mere economic value; I mean value in the philosophical sense.

For Naess and Leopold, ethics is largely about what warrants our care and consideration, and how we should behave toward those entities that demand this care. At the end of the day, we care about the things we love. And lovers can see and know things that others cannot. "Love is in fact a way of knowing, but its dynamics are the reverse of the usual models. Love comes first, and opens up possibilities" (Cheney and Weston 1999: 118). Declaring what you care about is not "soft," it is honest.

Having promoted the epistemological importance, or the "knowing" dimensions, of feelings, care, and love, we are confronted with one of the quandaries of present society. That is, how should we interpret the apparently similar signs of consciousness, care, and empathy observed in animals without falling into the "heresy" of anthropomorphism? But, what exactly is the problem with anthropomorphism? Is this a label wielded excessively by those who wish to see the world "objectively," as a way of dismissing those who are attempting to live more intimately in it? We are constantly imagining the world through the eyes of others - our friends, families, and lovers. This is one way that we come to understand them. It seems a short empathetic step to imagine the world through the eyes of more-than-humans.

Cultural texts about "talking middle class bears, and ducks" can clearly be criticized. However, how should we interpret Aldo Leopold's account of the green fire leaving the eyes of a dying wolf? More recently, Barbara Gowdy's book *The White Bone* (1998), a fictionalized account written from the perspective of an elephant, evoked much understanding and plausible empathy. Surely it is time to take these, and other expressions of care seriously.

If we are to take the advice of Leopold and Naess seriously, then education will be found at the edge between past practices and future possibilities. It will be looking beyond the status quo, social conventions, and petty rationality. But can we create learning opportunities that take us to this edge in ways that are educational? What would this look like? In what follows, I will lay out a few tentative guideposts that may assist educators in mapping the interiors of their curricular maps.

GUIDEPOSTS

Going forward toward new possibilities for environmental education is messy; education is itself a messy business. It requires constant reflection and examination on the part of practitioners. However, messy or not, practitioners must decide what kind of education they feel is needed and what this means for environmental education. We have to assume some responsibility for laying out our own guides to quality environmental education. We have to act; we have to create learning opportunities.

Each person must do some of the hard work for him or herself. However, expanding on earlier work (Jickling 2003), some tentative guideposts are offered; perhaps some will be useful.

Tell stories

Life is a story and it is shaped by countless other stories. Even science is a story played out within a set of assumptions. Thomas Kuhn (1970) made this much clear in his classic book, *The Structure of Scientific Revolutions*, where he illustrated how science does not exist apart from other ways of understanding the world. It is not objectively aloof and disinterested in the world.

As educators, we can enable students to explore those assumptions by asking: "What kinds of questions are being asked?" "Why?" and "How do the scientists seem to be approaching the world – as an object for experimentation, or as a subject of respect?" And, does this science help us to live a better – more just, equitable, and respectful – life. Answering these kinds of questions can enable us to be more conscious of the stories that we need.

We can also seek alternative stories to tell. Daniel Quinn (1992) does this in his novel *Ishmael* where, using a Socratic dialog between a gorilla and a man, he excavates many fundamental assumptions guiding Western societies. But there are many other stories; look for them. Seek them in your community and from other cultures, and tell them as if they were mirrors that help you and your students look back on yourselves, your values, and your culture.

Tell the story of the research subject described earlier in this paper. Embrace his love of natural history. But also make room for those stories being told by the social critics and non-conformists – the so-called "radicals."

Embrace ambiguity

When pursuing difficult questions about "a good life" and "living well," there isn't a right answer and there is always uncertainty. There is also uncertainty when we tackle contentious issues, and this leads to ambiguities. But, ambiguity should not paralyze or confuse; rather, it should invite tentativeness. It acknowledges multiple realities and multiple truths and it also allows you to be uncertain, or even wrong. It gives everyone room to move and to explore new possibilities.

Ambiguity should create intellectual and practical space for creativity, developing new ideas, new emotions, and stretching our ways of thinking and being. Ambiguity should create space to move beyond just sustainable development – to allow room for Greenpeace and other "radicals," and room to seriously consider Gainists, followers of the Deep Ecology movement, bioregional practitioners, ecofeminists, and the yet to be expressed formulations of ecological being.

Build in indeterminacy

As our practices become more advocacy-oriented they become more loaded, more prescriptive. And, as our practices become more loaded they become more adamant, more confident, more sure of themselves. Sometimes they become less tolerant of divergent ideas. When this happens we can become persuasive, even coercive, as we drift from educative aims toward more doctrinaire stances.

While our educational practices are never value-free, building in indeterminacy can be part of a conscious strategy to allow space for students to move beyond our interests and our prescription, and opportunities to disagree and to explore divergent opinions.

Be fair

For the Norwegian philosopher Arne Naess (2000) it is okay to pursue your particular environmental goals and to invite your students to do the same, but not in a wrong way. For him, leading students in a particular direction is always qualified through use of the word "if." "If you have," for example, "the following value priority, then [this relevant action follows]." It is also very important to respect alternative positions. It is appropriate to elaborate your own position, but not to end with that. You may end by clarifying the position of opponents. For Naess it is important to pursue truth and validity. "He would not claim an ultimate answer, but he would like to point in a direction" (Naess 2000: 62).

Don't end with your view; finish your discussion with views from other perspectives on the issue. This will help to address the concerns of your critics, but it may also allow your students to extend their reach beyond your limits. Just maybe, a Greenpeace member or other so-called "radical" will have something useful to contribute. It also means that there are economic and industry perspectives that need to be considered. This may not mean just allowing room for multiple perspectives, it may require actively seeking them.

Perform well

What you do counts; education has performative dimensions. It is more difficult, for example, to enable understanding and competency in democratic principles and actions if you run authoritarian classes. It just doesn't work well when educators work against themselves by espousing one set of values yet undermine them with their actions. Similarly, it is difficult to imagine an educational setting that is respectful of all life forms where students never see, touch, smell, or listen to other beings, or worse, settings where dissections are their only contact with organisms. All of those hundreds of little activities that comprise the daily life of an educator suggests an implicit curriculum that is loaded with values (Eisner 1985).

Be a citizen too

Educators should continue to have an active role in the affairs of their community. In fact, to do otherwise can carry the message that citizen participation is not important, that citizens cannot make a difference. At an increasingly cynical juncture in our history, students find it evermore important to see mentors "walking their talk." However, it is important to know when our actions, and influence, get in the way. Students can be impressionable, and coercion, however subtle or unintended, is not educational. They respect your involvement and your courage to take a stance on an issue, but they also want room to make their own judgments.

Select issues carefully

Our issues aren't necessarily our students' issues and what is important to them may not be as obvious as we think. If we move away from the student-centered program suggested by Simmons (1996) we might risk loading instruction with our agenda. This loading can be leading to our students and a subtle form of coercion.

Instead, begin by asking students to imagine how they might make a contribution to their community, how they might make a difference socially and environmentally. Then have them research issues and service-oriented projects where they can enact their contribution.

Value controversy

Issues can be complex and messy, but get involved anyway. A vibrant democracy depends on this participation, which is the very expression of discomfort and controversy. This is how we enact social critique and how we reveal assumptions and introduce new ideas (Thomashow 1989; Clarke 1992–93).

There are many ways to become involved. We may take part in a political forum or we may seek to reveal missing pieces, or silences, in the implicit curricula (Eisner 1985). However, careful preparation is required; success and failure can be separated by a heartbeat. The greater the controversy, the greater the need to present clear, explicit, and defensible educational theory and pedagogy.

Nurture feelings and care

Feelings are important, yet are often blank spaces on our curricular maps. As Leopold (1966: 251) said: "We can be ethical only in relation to something we can see, feel, understand, love, or otherwise have faith in." We need to pay attention to this.

Also, as Arne Naess (2000: 53) says: "There is an underestimation of the cognitive values of feelings." Following Naess, we should allow space in our instructional programs for his sequences of questions (Naess and Jickling 2000: 56): "How do you feel?" "What do you feel?" Then, "What should you feel?" "What do you think you are right to feel?" and "What do you want yourself to feel?" Naess' views are nicely complemented by Val Plumwood's (1999: 75) observation that "[m]oral reasoning requires some version of empathy, putting ourselves in the other's place, seeing the world to some degree from the perspective of the other with needs and experiences both similar to and different from our own." Building our ethical understanding will, therefore, be linked to our ability to develop emotional understanding. It is here that care and respect are grounded.

But seeing, feeling, understanding, caring, and loving are active responses to real relationships. This requires real contact with other people, social groups, societies, and more-than-human living beings. It requires contact with the land. The community and the earth become the textbook.

Be courageous

Good education that can enable change, that can transcend the status quo, requires risk. Take some chances. Some of the best education will take place on the edge between present realities and future possibilities. Perhaps it also takes place at that frustratingly movable line where education and advocacy finds an uneasy balance. Unfortunately, that place doesn't stay still. It moves around as issues and actors change. Good teachers will make some mistakes and will, from time to time, have to pull back. However, they will also be pushing the pedagogical and theoretical "envelope."

Advocate - in an educational way

In the years that followed my first encounters with local controversies, I did advocate vigorously on behalf of wolves, as a citizen but not as an educator. Mine was persistent and determined advocacy on behalf of a particular perspective, and it didn't include much room for a fair-minded representation of all points of view. The other sides were doing that for themselves. In a small and politically charged community this "wolf loving" wouldn't have struck anyone as educational.

When wearing my educator hat as a college instructor, I felt vulnerable to complaints about "brainwashing." If classes in environmental ethics were encouraging students to investigate philosophical questions about the human-nature relationship, I would have lost credibility by using the class as a platform for reinforcing my public views about wolves. College students are sensitive about this kind of opportunism and rebel against coercion. So, it was a delicate business to talk about wolves at the height of the conflict and in the heat of high running passions.

In hindsight, though, some of our collective efforts on behalf of wolves do seem to have had an educational effect. If education enables social critique, reveals hidden assumptions for public discussion, and disrupts the status quo, then citizens who spoke on behalf of wolves certainly did that. There was a vigorous public debate. And many community members gained confidence in their non-conformist positions. In thinking about it now, many press releases and commentaries were peppered with questions for the reader. Over and over the readers were challenged to probe their own value systems, and asked how they felt about particular actions. This too has educational merit. Even when acting as strong advocates, we can choose to make our case in ways that can leave an educational legacy.

CONCLUSION

In the end, our job is to tell good stories, but also to live good stories. Students notice. They seek authenticity and integrity in those they choose as mentors; they expect a lot from teachers, in their schools *and* communities.

Good storytellers select stories carefully and shift emphases and nuances in accordance with times and places – they will adapt them to fill a niche, or to fit into a particular context. Times change and so do the demands on educators. The equilibrium between education and advocacy is not static.

In my own story, the politically charged atmosphere of the Yukon wolf kill demanded that more attention be paid to educational integrity. It was important that my public agenda did not pre-empt educational opportunities, that my students had the intellectual space to think about their own values and to disagree, if they wished, with the positions that I had publicly declared. Now, some years later, and in a calmer political climate, emphasis has shifted toward integrity as a citizen. Educators, as mentors, earn regard and collegiality through exemplary actions, courage, commitment, care, and passion.

The stories of good teachers are often tales of a struggle between the impartiality expected in educational settings and their commitment as citizens. In the end, the relationship between education and advocacy remains a difficult one. Yet, educators are given responsibilities for guiding learning in spite of inevitable uncertainties and difficulties. To put aside controversial issues and radical questions risks conveying the implicit message that these ideas are not important, are not ones that learners should think about, and that learners cannot make a difference in the outcomes of these issues. On the other hand, if we provide students with the competencies to engage thoughtfully in critical issues and to take citizen actions, and if we exemplify these qualities ourselves, then we will increase their educational opportunities. And, in doing so, we will increase their space for human freedom. To do less would be to abrogate our responsibilities.[2]

NOTES

1. In the intervening years many of these authors are keen to note that their work has changed. See, for example, Cheak *et al.* (2002) and their attention to qualitative methodologies.

2. Similar expressions, and further explorations, of these ideas can be found in Kelly's (2001) ideas about "committed impartiality," and Jensen and Schnack's (1997) and Fontes (2004) ideas about action competence.

REFERENCES

Bacon, F. (1620). *The New Organon*, Book 1, Aphorism XCVIII. Republished in 1960. In *The New Organon and Related Writings*, ed. F. H. Anderson. Indianapolis: The Bobbs Merril Company, pp. 94–95.

Cheak, M., Volk, T., and Hungerford, H. (2002). *Molokai: An Investment in Children, the Community, and the Environment*. Carbondale, IL: The Center for Instruction, Staff Development and Evaluation.

Cheney, J. and Weston, A. (1999). Environmental ethics as environmental etiquette: toward an ethics-based epistemology. *Environmental Ethics*, **21**(2), 115–134.

Clarke, P. (1992–93). Teaching controversial issues. *Green Teacher*, **31**, 9–12.

Eisner, E. (1985). *The Educational Imagination: On the Design and Evaluation of School Programs*, 2nd edn. New York: Macmillan.

Evernden, N. (1985). *The Natural Alien: Humankind and Environment*. Toronto: University of Toronto Press.

Fien, J. (1993). *Education for the Environment: Critical Curriculum Theorising and Environmental Education*. Geelong: Deakin University Press.

Fontes, P. (2004). Action competence as a unifying objective for environmental education. *Canadian Journal of Environmental Education*, **9**, 148–162.

Gilbert, S. (1994). Science, ethics and ecosystems. In *Northern Protected Areas and Wilderness*, ed. J. Peepre and B. Jickling. Whitehorse: Canadian Parks and Wilderness Society and Yukon College, pp. 195–201.

Government of Canada. (2002). *A Framework for Environmental Learning and Sustainability in Canada*. Ottawa: Government of Canada

Gowdy, B. (1998). *The White Bone*. Toronto: Harper Collins.

Hines, J. M., Hungerford, H. R., and Tomera, A. N. (1986–87). Analysis and synthesis of research on responsible environmental behavior: a meta-analysis. *Journal of Environmental Education*, **18**(2), 1–8.

Hopkins, C. (1998). *Environment and society: education and public awareness for sustainability*. Proceedings of the Thessaloniki International Conference organized by UNESCO and the Government of Greece (8–12 December 1997), pp. 169–172.

Hungerford, H. R., Peyton, R. B., and Wilke, R. J. (1980). Goals for curriculum development in environmental education. *Journal of Environmental Education*, **11**(3), 42–47.

Jensen, B. B. and Schnack, K. (1997). The action competence approach in environmental education. *Environmental Education Research*, **3**(2), 163–178.

Jickling, B. (1991). Environmental education and environmental advocacy: the need for a proper distinction. In *To See Ourselves/To Save Ourselves: Ecology and Culture in Canada*, ed. L. A. Iozzi and C. L. Shepard. Montreal: Association for Canadian Studies. pp. 169–176. (An earlier version of this paper appeared in L. A. Iozzi and C. L., Shepard, ed. *Building Multicultural Webs through Environmental Education*. Troy, OH: North American Association for Environmental Education, pp. 143–146.)

(1992). Why I don't want my children to be educated for sustainable development. *Journal of Environmental Education*, **23**(4), 5–8.

(2003). Environmental education and environmental advocacy: revisited. *Journal of Environmental Education*, **34**(2), 20–27.

Jickling, B. and Spork, H. (1998). Education for the environment: a critique. *Environmental Education Research*, **4**(3), 309–327.

Kelley, T. (2001). Discussing controversial issues: Four perspectives on the teacher's role. In *Philosophy of Education*, 3rd edn., ed. W. Hare and J. P. Portelli. Calgary: Detselig Enterprises Ltd., pp. 221–242.

Kuhn, T. (1970). *The Structure of Scientific Revolutions*, 2nd edn. Chicago, IL: University of Chicago Press.

Leal Filho, W. (2003). *Towards a Modernization of Environmental Education*. Presentation at the First World Environmental Education Congress, Espinho, Portugal, May 20–24.

Leopold, A. (1966). *A Sand County Almanac: With Essays on Conservation from Round River*. New York: Sierra Club/Ballantine. (First published in 1949/53).

McKeown, R. and Hopkins, C. (2003). EE ≠ ESD: defusing the worry. *Environmental Education Research*, **9**(1), 117–128.

Naess, A. (2002). *Life's Philosophy: Reason and Feeling in a Deeper World*. Athens, GA: University of Georgia Press.

Naess, A. and Jickling, B. (2000). Deep ecology and education: a conversation with Arne Naess. *Canadian Journal of Environmental Education*, **5**, 48–62.

Peters, R. S. (1973). The justification of education. In *The Philosophy of Education*, ed. R. S. Peters. Oxford: Oxford University Press, pp. 239–267.

Plumwood, V. (1999). Paths beyond human-centeredness: lessons from liberation struggles. In *An Invitation to Environmental Philosophy*, ed. A. Weston. New York: Oxford University Press, pp. 69–105.

Pratt, D. (1980). *Curriculum Design and Development*. San Diego: Harcourt Brace Jovanovich.

Quinn, D. (1992). *Ishmael: An Adventure of the Mind and Spirit*. New York: Bantam/Turner.

Sanera, M. (1998). Environmental education: promise and performance. *Canadian Journal of Environmental Education*, **3**, 9–26.

Sanera, M. and Shaw, J. (1996). *Facts not Fear: A Parent's Guide to Teaching Children About the Environment*. Washington, DC: Regnery.

Saul, J. R. (1995). *The Unconscious Civilization*. Concord, Ont.: Anasi.

Scheffler, I. (1960). *The Language of Education*. Springfield, IL: Charles C. Thomas.

Schumacher, E. F. (1973). *Small is Beautiful*. New York, NY: Harper & Row.

(1977). *A Guide for the Perplexed*. New York, NY: Harper Colophon.

Scoullos, M. J., ed. (1988). *Environment and Society: Education and Public Awareness for Sustainability*. Proceedings of the Thessaloniki International conference organized by UNESCO and the Government of Greece (8–12 December 1997). Athens, University of Athens, MIO-ECSDE & Ministry for the Environment, Ministry of Education.

Shaull, R. (1970). Foreword. In *Pedagogy of the Oppressed*, P. Freire. New York: Continuum, pp. 9–15.

Sia, A. P., Hungerford, H. R., and Tomera, A. N. (1985–86). Selected predictors of responsible environmental behavior: an analysis. *Journal of Environmental Education*, **17**(2), 31–40.

Simmons, B. (1996). President's message. *Environmental Communicator*, **26**(4), 2–3.

Stapp, W. B., Bennett, D., Bryan, W., Fulton, J., MacGregor, J., Nowak, P., Swan, J., Wall, R., and Havlick, S. (1969). The concept of environmental education. *Journal of Environmental Education*, **1**(1), 30–31.

Thomashow, M. (1989). The virtues of controversy. *Bulletin of Science, Technology & Society*, **9**, 66–70.
UNCED. (1992). *Agenda 21: The United Nations Program of Action from Rio*. New York: UN Department of Public Information
Weston, A. (1992). Before environmental ethics. *Environmental Ethics*, **14**(4), 321–338.
World Commission on Environment and Development (WCED). (1987). *Our Common Future*. Oxford: Oxford University Press.

JOY A. PALMER AND JOANNA C. BIRCH

6

Changing academic perspectives in environmental education research and practice: progress and promise

INTRODUCTION

Environmental education research and practice have developed and changed significantly over the past half-century. Approaches to environmental education have changed from a "nature study" and natural science base of the 1960s and early 1970s to an action research and social sciences-orientated perspective of the 1990s and present day. Academic perspectives regarding the aims for environmental education and the role of the natural and social sciences in environmental education have also become increasingly diverse and complex. This chapter provides an overview on the changing approaches to environmental education practice and some of the inconsistencies between practice and academic perspectives for environmental education. The chapter also provides an overview on changing perspectives within environmental education research and concludes with suggestions on how advances in research, from the early 1990s to 2000s, might best inform environmental education practice.

CHANGING APPROACHES IN ENVIRONMENTAL EDUCATION PRACTICE

Education is far from realizing its maximum potential of helping people understand and appreciate the environment and their role with respect to sustainable development. This is perhaps a surprising statement given researchers' and educators' engagement with the growth and development of diverse approaches to environmental and related "educations" over the past four decades (Figure 6.1). Approaches to environmental education have moved from the nature studies and fieldwork activities of the 1960s, with an academic focus on

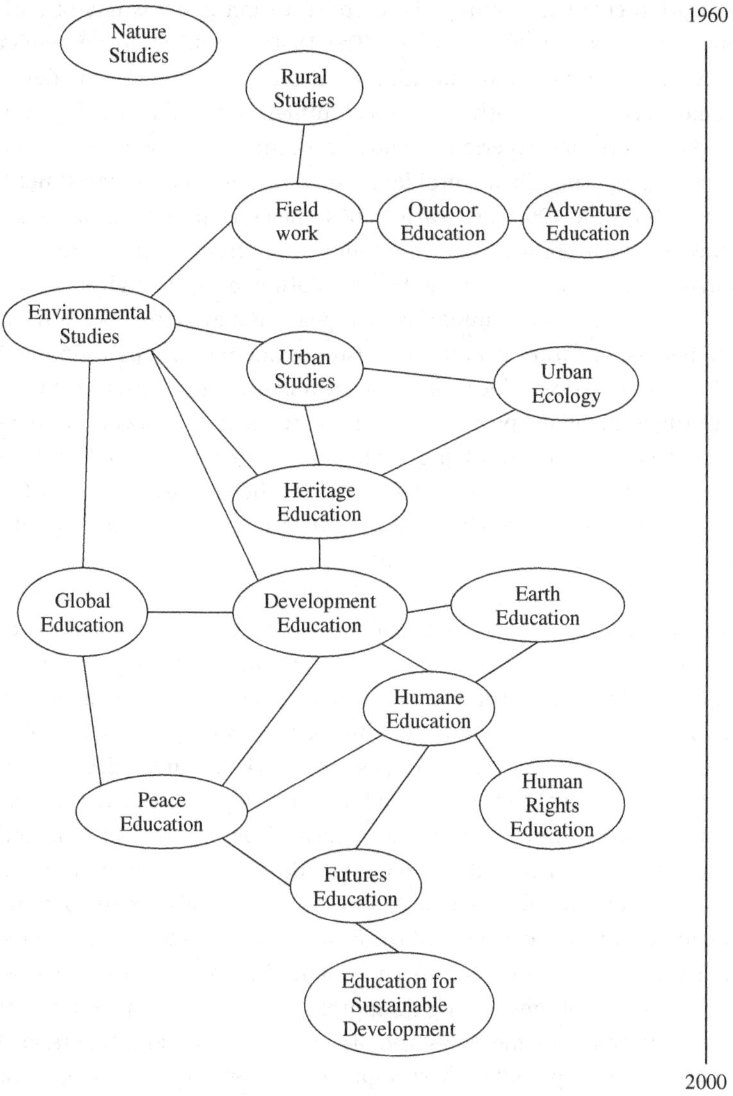

Figure 6.1. Development of different aspects or emphases of environmental education from 1960 to 2000 (from Palmer 1998a: 27, Figure 1.3).

biology or ecology, to the emergence of the environmental studies, urban studies, and conservation/heritage education of the 1970s. This decade was characterized by increased use of the natural environment for first-hand experiences, the growth of field and environmental/outdoor education centers, teaching about conservation issues, study of built environments and urban ecology. The 1980s heralded an even wider vision of environmental education to include global education, development education with a political dimension, value clarification through personal experience, and community-based action research encouraging pupil-led problem solving *in* the environment. Throughout the 1990s and early 2000s, the concept of empowerment emerged, incorporating capacity building, communication, problem-solving and action aimed at the resolution of socioenvironmental problems. A focus on human rights, peace issues, and education for the benefit of future generations also became part of environmental education practice. Education for a sustainable future emerged in the late 1990s, involving participatory action research approaches to changing behaviors and resolving ecological problems. The notion of a global community, wherein students, teachers, researchers, NGOs, and politicians around the world work together to identify and resolve socioecological problems also began to emerge.

All of these approaches and trends in related "educations" have both *contributed* to and *conflicted* with the steady evolution of environmental education and education for a sustainable future (Palmer 1998a). The approaches have a great deal in common and their aims, objectives, and content merge and overlap to a considerable degree. Nature studies, rural studies, fieldwork, urban studies, conservation and a host of "adjectival" "educations" (e.g., outdoor, adventure, earth, values, humane, heritage, peace, development, global, and futures) have contributed to growth and development of environmental education practice. Inevitably, however, the existence of so many related yet distinct elements of the evolving field has led to fragmentation, conflict, and numerous instances of needless reinventing of wheels.

A review of these changing approaches also reveals a sequence of major international meetings, initiatives, and publications that helped shape the development of these approaches. The speed of development of thinking and documentation related to environmental education in the 1970s and early 1980s was quite remarkable and resulted in foundational policy statements on environmental education, such as The Belgrade Charter: A Global Framework for Environmental Education (1975), the Tbilisi Declaration (1977 IUCN), and World Conservation Strategy (1980). The late 1980s and 1990s saw no reduction in the number of international conferences, published statements of definition, aims and goals, and national guidelines on teaching and learning in relation to the environment. At start of the twenty-first century, international acknowledgment emerged at all levels, through formal documents published by many governments (e.g., UK Government Panel for Sustainable Development Education) on the importance of environmental education, or education for sustainable development, and the entitlement of all to receive it. Environmental education researchers and educators have witnessed a substantial refinement of the language used to talk about environmental education in these influential documents and in the development of approaches to environmental education. But what is the effect of all this upon everyday practice?

As stated above, environmental education still appears to be a long way from reaching its potential. There still appears to be a substantive gap between actual practice and academic perspectives presented in international conferences, documents, and guidelines, referred to as a "Rhetoric–Reality Gap" in Palmer (1998a: 96). The gap leads to conflicts, inconsistencies, and limitations in environmental education practice.

For example, Stevenson (1987) outlines a series of major contradictions between environmental education documents and actual practice. The goals, principles, and guidelines of environmental education presented in UNESCO (1975, 1977), or IUCN, UNEP, WWF (1991), documents suggest particular curriculum orientations and teaching practices that engage students in problem-solving and action-based activities with a focus on real environmental issues. These approaches call for interdisciplinary and flexible inquiry. For some authors, the introduction of these approaches into a school curriculum represents a fundamental challenge to the dominant conception of and approach to the organization and transmission of knowledge (Esland 1971). Robottom (1982, 1983) and Volk *et al.* (1984) also elaborate on the discrepancy between the acquisition of environmental knowledge and awareness in "traditional" school programs, and the action-orientated goals of the contemporary rhetoric of environmental education within policy statements and research. These authors, and others, contend that school curricula tend to be discipline-based and emphasize abstract theoretical problems. For example, Stevenson (1987: 75) states:

> The common curriculum emphasis can be described as the mastery of many fragmented facts, concepts and simple generalizations organised loosely within discrete bodies or fields of study. The predominant pedagogical process involves the teacher as dispenser of factual knowledge . . .

In the environmental education policy statements, students are to become active thinkers and generators of knowledge, whereas in schools students are usually perceived to be in a passive position as spectators and recipients of other people's knowledge and thinking (Stevenson 1987).

Another example of the mismatch between the rhetoric and practice relates to the breadth of coverage or primary location for environmental education within subject-based curricula. Generally accepted policy and guidelines expect environmental teaching and learning to be both interdisciplinary and cross-curricular. Environmental education is viewed as an approach to education rather than as a subject. Yet in practice there is a very strong emphasis on grounding environmental education within science or geography curricula. This situation is very apparent in many locations around the world. Programs of study in science and geography ensure that environmental matters are taught, and schools may choose to include environmental education in other National Curriculum subjects. However, environmental education only extends beyond the scientific domain if the motivation and enthusiasm of individual teachers and schools so decide.

Perhaps the solution, and root, of such inconsistencies, conflicts, and limitations (the "Rhetoric–Reality Gap") may lie in the changing academic perspectives on environmental education research. For example, Fien (1992), Elliot (1991), and Robottom and Hart (1993), amongst others, challenge the dominant scientific paradigm within environmental education. Consequently, there has been a major shift in perspectives regarding environmental education research throughout the 1990s.

CHANGING PERSPECTIVES IN ENVIRONMENTAL EDUCATION RESEARCH

There are three key perspectives within environmental education research during the 1990s and early 2000s that have influenced the relationship between research and actual practice in environmental education. The perspectives are: (a) quantitative scientific research studies still tend to dominate environmental education research; (b) the number of qualitative (including interpretive, constructivist, and socially critical) research studies has increased significantly; and (c) there is an increased emphasis in research on grounding methodological positions in practice and linking improvement of education practice with empirical research.

Quantitative research perspectives

The dominant scientific worldview appears to have had an overwhelmingly powerful influence not only upon the practice of environmental education, but also upon the development of research. The development of an environmental education research agenda gained momentum in the United States during the late 1970s and the 1980s. Leading researchers of the time tended to be educationalists whose quantitative approaches dominated developments and published research outcomes (see, e.g., Fensham 1978; Lucas 1979; Stapp *et al.* 1980; Hungerford *et al.* 1980, 1983; Hungerford and Volk 1990). Marcinkowski (1990) provides an overview of this approach, wherein research approaches are derived from the natural, physical, and social sciences, reflecting the tradition of scientific inquiry. The aims of research are to establish patterns of, relationships between, and causes of social phenomena by way of description, prediction, and explanation. Prototypical designs include survey, correlational, and experimental approaches, with results presented in numerical and statistical form. In this approach the researcher remains detached from the research setting to avoid bias and tends to rely upon instruments as an intermediary device for data collection purposes.

The majority of the research published in the 1970s and 1980s in the *Journal of Environmental Education* reflects these characteristics. Many studies of the period were concerned with the identification, prediction, and control of the variables that are believed to be the critical cognitive and affective determinants of responsible environmental behavior. Iozzi (1981) reports that 90–92 percent of research studies in environmental education at this time were quantitative. Substantive critiques and international debate of such quantitative approaches emerged during the early 1990s.

Qualitative research perspectives

There was a call for a broadening of research approaches and a shift toward the use of more humanistic and interpretive lines of inquiry (see, for example, Robottom and Hart 1993; Palmer 1998a; Elliot 1991). Such a broadening and shift gained momentum during the mid-1990s, to include *interpretivist* (constructivist) and *critical* research paradigms.

The *interpretivist* paradigm focuses on understanding the subjective world of human experience. For example, it could focus on investigating how students and teachers conceptualize the environment, how people develop the meaning of environmental concepts, individuals' reflections and experiences related to the environment, and studies on the use of language in environmental teaching and learning. It is grounded in interactive, field-based inductive methodology. Research questions may be explored through interpretive modes of inquiry, such as case studies, participant observation, semi-structured interviews, discourse analysis, connoisseurship, and criticism. The term *constructivism* is sometimes used synonymously with, or alongside, *interpretivism* in descriptions of this type of qualitative research.

The second research perspective, which emerged from and shares the interpretivist critique of the quantitative scientific approach, is the *critical theory* or *socially critical* approach. Advocates of the critical paradigm argue that our subjective views are not only internally constructed but also influenced by persuasive social forces. Individuals or groups cannot be considered separately from their social context. Robottom and Hart (1993) provide a useful account of the "socially critical" approach to environmental education, which should ideally involve students, teachers, and community agencies in collaborative investigations of real environmental issues in their local environments. The investigations should be socially critical in that they seek to uncover and make explicit the values and vested interests of the individuals and groups who adopt positions with respect to the issue. Three key forms of data collection and analysis employed by educational researchers in the critical tradition include discourse analysis, critical ethnography, and action research. Fien and Hillcoat (1996) provide a useful introduction to these approaches.

Current research perspectives

Although environmental education research has moved away from its roots in the scientific paradigm toward a broader base of qualitative methodologies, the quantitative approaches continue to be a powerful influence in environmental education research and practice. Rickinson (2001) provides a comprehensive review of a large body of environmental education research published between 1993 and 1999 that demonstrates that most evidence about environmental learning and learning in general, collected during the latter half of the 1990s, is still quantitative. Rickinson (2001: 302) noted: "there is more information about learners than learning, more information about students' environmental characteristics than their educational characteristics and more information about learning outcomes than processes." Environmental education researchers are being encouraged to pursue more rigorous and scientific forms of qualitative methodology, as the qualitative research of the 1990s is criticized for its small samples and lack of control groups (Rickinson 2001). Many of the qualitative methodologies were perceived as lacking in rigor or poorly articulated, thus raising serious questions about reliability and validity.

Current environmental education research is confronted with the task of building upon areas of research carried out during the 1970s and 1980s and designing educational programs that recognize differences between learners, fit a variety of needs, relationships, and settings. Environmental educators are encouraged to respond to the fact that: "Students are active experiencers, rather than passive recipients, of environmental curricula and respond to different learning situations in different ways" (Rickinson 2001: 302). Such "tailoring" has long been suggested by environmental education researchers (Arbuthnot 1977; Hines *et al.* 1986–87; Juniper 1989; Arcury and Christianson 1993; Scott and Willits 1994) and it remains obvious that we need to research and practice environmental education so that it reflects understanding and acknowledgment of educational settings and relationships such as those between ethnicity, culture, social class, and gender (Rickinson 2001).

However, environmental education researchers are challenged to avoid repeating the quantitative, scientific research that dominated the field's early years. Researchers should no longer be expected to identify clear socio-demographic predictors of environmental knowledge, concern, or behavior. Recent research practices have indeed embraced more diverse aspects of environmental understandings along with their social and physical contexts. The research has focused increasingly on environmental attitudes and the processes of thinking and learning, and rather less upon knowledge outcomes, action, links between knowledge, action and behavior, or predictors of action. There is also an increased emphasis being placed on linking empirical research to the improvement of educational practice.

APPLICATION OF RESEARCH TO PRACTICE: PROGRESS AND PROMISE

There are four examples of research from the early 1990s to the present that demonstrate an obvious application to practice and enable researchers and educators to close the *"rhetoric-reality"* mismatch between academic perspectives and educational practice. The four aspects of recently reported research chosen as our examples are: (a) the importance of formal education as a significant influence upon environmental thinking and behavior; (b) learners' knowledge and understanding of environmental issues; (c) the significance of outdoor and fieldwork experiences; and (d) interpersonal relationships as an integral part of environmental learning. These four aspects of research have been studied recurrently over the last 30 years and still stand as key contemporary topics of investigation. There is also evidence that these aspects of environmental education research are currently being applied successfully to environmental education practice.

The importance of formal education

Both research and global environmental initiatives have repeatedly endorsed the importance of formal education as a means of imbuing learners with sound environmental knowledge and achieving behavioral change. The use of formal education as an effective environmental education approach has received continued recommendation in international policy and education initiatives such as the *Caring for the Earth* (IUCN, UNEP, WWF 1991) and UNCED's (1992) follow-up *Agenda 21*. Throughout the 1990s, various governments have interpreted the environmental initiatives by preparing publications and curriculum guidance for environmental education within formal education settings.

There is a substantial body of research that has sought to understand more about the role of formative influences and life experiences of learners on environmental learning. Chawla (1998) presents a review of this research into life experiences as sources of positive environmental attitudes. The instigator of this research field is recognized to be Tanner (1980). During the 1990s, Palmer and her colleagues also engaged in work with samples larger than many of the previous and subsequent studies (Palmer 1993, 1998a; Palmer and Suggate 1996). The work of Horwitz (1996), Sward (1996), Kidd and Kidd (1997), and Chawla (1998, 1999) extended this field of study across a number of age groups and cultures. This body of research reveals that the following key life experiences influence environmental learning and attitudes: (a) outdoor experiences; (b) the age or phase during which experiences occur, most notably childhood experiences of nature; (c) formal and informal education; (d) interpersonal relationships; (e) cultural influences; and (f) negative experiences. The following discussion will focus on how recent "life experience" research can help inform formal education practice.

Sward's (1996) findings reveal that only 12 percent of respondents cite teachers and peers as important influences. Kidd and Kidd's (1997) figures reveal that formal education in the form of science classes accounts for only 5 percent of the adolescent sample's interest in wildlife. Palmer (1993) points to the potential influence of formal education later in life when coupled with other experiences: "For some respondents, environmental concern developed from a childhood love of nature and the outdoors, followed by 'latent' teenage years, and then a refueling of enthusiasm while in higher education or on becoming a parent." Sward (1996) also highlights the importance of the later stages of formal education by proposing that secondary school may allow those who have not been exposed to experiences of nature or the outdoors to learn elements of environmental sensitivity.

It seems that, for some people, secondary school may provide some important influences that are more significant than primary school experiences but less influential than time spent in university or higher education (Palmer 1993). Other studies confirm that a large proportion of important life experiences (60 percent of reported responses) occur during either secondary or tertiary education (Palmer and Suggate 1996). However, this finding is not necessarily consistent around the world. Slovenian environmental educators (taking part in a nine-country study) cite particularly influential activities during primary education (Palmer *et al.* 1998). These data reveal that primary education *may* be important and that educational experiences of other cultures should be investigated further. From research undertaken so far, it can be concluded that the most significant experiences of childhood lie outside the primary-years formal education system.

The overall influence of formal education should not, however, be discounted. Palmer *et al.* (1999) find that education is one of four "leading groups" of experiences, along with "experiences of nature," "people," and "work." Sward (1996) ranks formal education third amongst other influences mentioned by her sample of El Salvadoran environmental professionals. Chawla (1998) cites Gundersons' (1989) study, which demonstrates 83 percent of respondents citing the important influence of both time spent outdoors and former teachers. Horwitz (1996) classifies influential experiences cited by her sample as "education" to include learning biology, ecology, natural history, philosophy, field trips at high school, outdoor areas, Youth Conservation Corps, reading, and teaching others.

Evidence concerning how such research transfers to formal education practice is difficult to assess, but there are clear indications of countries and communities engaging with such research agenda. There are numerous positive examples of countries that are making a bold response to *Agenda 21* using this research. Sterling (2003) refers to 11 Baltic countries planning to shift their entire formal education systems towards Sustainability. In the UK, we can also report Bloomfield's (1998) positive example of the adoption of *Agenda 21* issues within formal education settings. At the University of Hertfordshire, action workshops for teachers and student teachers on *Agenda 21* as a cross-curricular topic demonstrate the transference of research findings about tertiary education into practice. Both Sterling's (2003) and Bloomfield's (1998) references to the permeation of the themes of *Agenda 21* into formal education infer the encouragement of social action and the sharing of responsibility for environmental issues across the community, advocated by proponents of socially critical research. Student teachers following university courses are presented with the opportunity to "refuel" earlier enthusiasm for the environment which research has shown to be significant (Palmer *et al.* 1999) or consolidate knowledge developed during earlier years' education (Caro *et al.* 1994).

In Slovenia, various research and curriculum development projects have made a significant impact upon policy for and teaching of environmental education. Data from the Emergent Environmentalism project (Palmer 1999) have been widely disseminated to ministry personnel and practitioners through publications, seminars, and courses for teachers. Educational legislation, introduced in 1996, incorporates specific references to nature education and environmental protection and it can be expected that those involved in preparing new teaching programs will necessarily define basic targets for nature education and protection within these programs.

Learners' knowledge and understanding of environmental issues

There is an assumed fundamental connection between formal education and the transmission of environmental knowledge. As stated earlier in the chapter, teachers have traditionally been viewed as distributors of knowledge. If environmental knowledge is most effectively transmitted through formal education, then there is indeed a case for environmental education to be included in the curriculum of schools, colleges, and universities. However, formal education *about* the environment (i.e., solely knowledge-based education) has received criticism substantiated by the findings of environmental education research (Huckle 1993; Robottom and Hart 1993; Palmer 1998a). Educational practitioners who are informed about recent research findings now recognize that awareness and knowledge of environmental issues alone are not sufficient to elicit positive environmental behavior (Hungerford and Volk 1990; Palmer 1995, 1998b, 1999).

Environmental educators are now better informed about the processes of gaining environmental knowledge through recent research on the content, understandings, and misunderstandings associated with learners' environmental knowledge. Rickinson (2001) identifies "environmental knowledge" as one of six key nodes of research evidence having an impact on learners and learning to emerge from his review of the 1993–99 environmental education research literature. Misunderstandings and low levels of knowledge of some environmental issues are apparent from the last decade's research into learners' environmental knowledge (Rickinson 2001). There is evidence that students' thinking can lack structure (Rickinson 2001) and that children as young as four years are observed to have various conceptions and misconceptions of environmental issues such as global warming, deforestation, and waste management (Palmer *et al.* 1996, 1999). The potential resilience of inaccurate ideas presents environmental educators with the need to recognize and respond to concepts of "environment" held by children during their first years of schooling. For young children of four to seven years, studies also illustrate the existence of basic environmental knowledge and a capacity for "sophisticated reasoning and thinking" (Palmer *et al.* 1996). For these reasons, there is a fundamental need for environmental educators to understand and respond to the progression of children's environmental understanding.

Such research as has informed us of the potential for *development* of environmental knowledge may be supplemented with findings from work aiming at understanding what happens to environmental knowledge later in the life of the learner. Caro *et al.* (1994) suggest that undergraduates' knowledge of conservation biology may have an effect upon altering their environmental attitudes. A "crystallization" of students' attitudes toward the natural world may be achieved through learning undergraduate course material. They consider students' "latency" of environmental attitudes to exist perhaps in the same way that children possess a latency of knowledge to be explored and developed. It is clear that we still need to know more about learners' changing and developing perspectives over time (Rickinson 2001).

In addition to highlighting the importance of knowledge itself, there is a clear need for researchers and practitioners to recognize children's articulation of and action in response to that knowledge. The call for environmental knowledge to include knowledge of *courses of action* has been indicated during earlier phases of research (Hines *et al.* 1986–87). Now such a call is justified by various studies demonstrating that education programs do not always express sufficiently to learners how and why certain environmental knowledge or behavior is relevant to them (Chipeniuk 1998; Gough 1999).

Similarly, educators' are now informed from research that young people may be less concerned with the environmental issues that relate to them personally and rather more concerned with global issues (Uzzell *et al.* 1995; Morris and Schagen 1996; Connell *et al.* 1998). Sward's (1996) work and other studies have highlighted people's experiences of and concern for local environmental degradation, hence providing contrasting evidence. Nevertheless, this evidence provides the same message: *environmental education content should include exploration of local issues and personally relevant environmental matters for learners.* Additionally, researchers and practitioners are encouraged to consider learners' existing perceptions of nature (see Keliher 1997; Bonnet and Williams 1998). Our awareness that learners can perceive nature in connection with conceptions of leisure and solitude, and at the same time consider nature as frightening or under threat (Rickinson 2001), should encourage deeper consideration of how we explore *processes* concerned with building and expressing such perceptions.

There remains a need for both researchers and practitioners to communicate known cases where *processes* of environmental learning lie at the heart of education practice. However, some current practices are characterized by a prevalence of "market/managerial/target" driven education (Sterling 2003). For example, reported changes to South African environmental education (Schreuder *et al.* 2002), although commendable in many respects, have been designed to concentrate upon an "outcomes-based education" system that appears to remain connected to past paradigms of education. By contrast, Hungarian environmental education practice reported by Csobod (2002) is characterized by school initiatives that focus on developing the "action competence of the learner in addressing environmental issues." Through a series of workshops aimed at building teachers' confidence to deliver flexible education materials, students gain *skills* for environmental action, not *just* knowledge, with their teachers in the role of facilitators rather than instructors.

The significance of outdoor and fieldwork experiences

Outdoor experiences and fieldwork exemplify learning approaches that have long been heralded by environmental education researchers as successful approaches in developing environmental concern and/or action (Tanner 1980; Peterson and Hungerford 1981; Gunderson 1989; Palmer 1993; Palmer and Suggate 1996; Sward 1996). Past and current research concerning outdoor experiences and fieldwork in environmental education offer direct evidence regarding processes of environmental learning that can be applied in practice. However, Sterling (2003) and others suggest that environmental education practice is still not sufficiently learner-centered, process, or experience-based.

The relevance of outdoor experiences as part of environmental education came to light in the 1980s (Tanner 1980; Peterson and Hungerford 1981; Votaw 1983; Gunderson 1989). Hungerford and Volk (1990: 14) stated the importance of any environmental program offering outdoor experiences, whilst inferring that the experiences of educators themselves are significant and such educators may form key interpersonal relationships with learners: "it seems important that learners have environmentally positive experiences in non-formal outdoor settings over long periods of time. And in the formal classroom, we must look to teachers who are themselves sensitive and willing to act as the role models for learners." It is unsurprising that Chawla's (1999) more recent research still recommends the use of outdoor experiences, both in and out of school, as a major part of any environmental education program.

More recent studies of the last decade, such as those carried out by Palmer and colleagues, also reveal more of the nature of these outdoor experiences and of those who cited their importance. Palmer and Suggate (1996) found that the older members of their sample, mostly males over 50 years, cite outdoor experiences as most influential. For this group, some outdoor activities appear to be more influential in later life, for example gardening and agriculture. In 1999, Palmer *et al.* substantiated earlier research by revealing outdoor experiences to be one of the four major groups of significant influences for environmental educators.

Chawla also acknowledges the importance of the *timing* of outdoor experiences in her 1999 study in which it seems that the development of environmental concern or action are greatly influenced by outdoor experiences during childhood or adolescence. Of Sward's (1996) sample, 100% had engaged in outdoor experiences during their youth. In Horwitz's (1996) study an attempt is made to identify experience and influences across generations, ranging through six stages: early childhood, middle childhood, adolescence, early, middle, and recent or ongoing adulthood.

A variety of recent international studies continue to investigate the influence of fieldwork in education, for example, Uzzell *et al.* (1995) from the UK, Emmons (1997) from Belize, Bogner (1998) from Germany and Dettmann-Easler and Pease (1999) from the mid-west United States. Discussion of schools' use of fieldwork is included in Palmer and Neals' (1994) writing on provision of experiential education. The importance of fieldwork for first hand experience of nature is regarded as vital for children, especially for those of primary school age. From the work of the National Association of Field Studies Centre Officers (NAFSCO), Palmer and Neal (1994) draw upon guidelines for fieldwork. It is suggested that fieldwork should provide students with a desire for action, involve investigation, include a balance of study between people and nature; and that fieldwork should be frequently evaluated in terms of its success. We are in agreement that fieldwork as outdoor experience, and as an opportunity to gain direct awareness of nature and environment, is clearly "a valuable point of entry, a rich source of illustration, a stimulus to action and an aspect of the ultimate reason for environmentalism" (Dettmann-Easler and Pease 1999).

A good example of how this research can help inform practice can be found in the educational work of a number of non-governmental organizations (NGOs). For example, The Wildlife Trusts, a UK non-governmental conservation organization, offers a variety of outdoor experiences across 47 county locations. This NGO is well placed to offer experience-based environmental education on wildlife sites and reserves (Palmer and Lancaster 1999). The Wildlife Trusts provides practical, outdoor-based environmental activities for the young through "Wildlife Watch" clubs. Such activities provide a good illustration of the convergence of research evidence and trends in practice.

If, as discussed previously, the most significant environmental experiences of childhood lie outside the formal education system, then an informal club setting with its investigative activities has the potential to provide particularly significant experiences. The National Riverwatch project is merited for two reasons. Hart (1997) praises it for the good survey pack it provides for young children and adults and for its emphasis upon involving people. Likewise, the 1994 Enviroscope program is praised by Hart for enabling children to experience and survey small-scale local environments. The Watch Ozone project of

1992 is lauded for its aptness in encouraging children to monitor changes in their own local environment. These Wildlife Trusts projects appear to present young people with outdoor and field-based experiences of their local environment whilst also adopting a socially critical approach that investigates real issues in local environments suggested by the research of Robottom and Hart (1993).

A second good example of the convergence of research and practice lies in the work of the Field Studies Council (FSC) from the UK. As an education charity with 17 centers, the FSC provides people of all ages with multifarious outdoor courses and experiences including walking, adventure, arts, photography, ecology courses, traditional use of natural materials, family wildlife experiences, and overseas outdoor experiences. The educational work of the Field Studies Council links educational research evidence to practice in that a *range* and *variety* of outdoor events and activities are offered. It has been shown that people of different ages respond to varying outdoor activity stimuli (Palmer and Suggate 1996) and that variation and tailoring of experience to suit people's environmental education needs is a positive element of practice (Arbuthnot 1977; Hines *et al.* 1986–87; Juniper 1989; Arcury and Christianson 1993; Scott and Willits 1994).

Outdoor and fieldwork activities are not confined to the practice of NGOs, but are also endorsed by government bodies. For example, the UK's Qualifications and Curriculum Authority suggests that schools could involve their pupils in tree planting and gardening, embracing the value of outdoor experience and the assistance of NGOs (QCA 2003). Overall, the implementation of activities provided in the outdoors, in reserves, and in school grounds, corresponds to Rickinson's (2001) conclusion that students become *active learners* and not passive recipients of environmental curricula and experience.

Interpersonal relationships

A final key learning experience revealed by recent environmental education research as significant and highly applicable to education practice concerns the influence of interpersonal relationships in developing environmental concern. The relevance of interpersonal relationships for environmental education practice lies in research evidence that illuminates how certain *people* can provide highly influential life experiences that may stimulate environmental concern and action. Research indicates the potentially highly significant roles of parents, NGO leaders, and other adults in environmental education. Furthermore, evidence from recent research into interpersonal relationships may also be considered a valuable means of enabling environmental education to be practiced more holistically than it has been in the past - blurring boundaries between formal and informal education, home and school, government, and non-government; knowledge and experience - permitting sustainability education to shed some of its "tag-on" status within schools, as lamented by Sterling (2003).

The notion that friends, parents, families, or other adults can play an important role in environmental education has been confirmed by the research of Palmer (1993, 1999), Horwitz (1996), and Chawla (1999). Kidd and Kidd (1997) find that adults act as role models for young adolescents. Interpersonal relationships may also develop through friendships at university (Chawla 1999), the influence of older friends, and having children (Palmer and Suggate 1996). During adulthood, worrying about and wanting to contribute to the well-being of future generations influence decision-making and pro-environmental disposition (Horwitz 1996; Chawla 1999). For this reason, Chawla (1999) recommends outreach to parents as an important aspect of environmental education. Hart (1997) suggests that school settings should also provide opportunities for parents to engage in the processes of environmental education.

Raising environmental awareness among parents as well as pupils is discussed in Rovira's (2000) appraisal of Catalonian environmental programs. Although Rovira concludes that it is doubtful that the education of parents through school projects is especially effective, the work of Ballantyne *et al.* (1998) offers an alternative perspective. Their work investigates what they refer to as "intergenerational learning" and they conclude that the potential effects of students' engagement in environmental education upon other members of the family cannot be ignored. The research of Uzzell *et al.* (1994) provides a European example to complement the work in Australia carried out by Ballantyne *et al.* (1998). Both studies inform us about the *processes* of intergenerational communication between pupils and adults. American and British studies from Leeming *et al.* (1997) and Evans *et al.* (1996) respectively provide us with information about *outcomes* of intergenerational environmental learning programs. As Rickinson's (2001) review informs us, young people do have the capacity to influence their parents' environmental behaviors, in particular through planning of programs which include adults as well as children, tackle local and relevant environmental issues, and enable young people to *enjoy* environmental education programs.

A good example of educational practice that encompasses such research findings on interpersonal relationships may be found in the UK Qualifications and Curriculum Authority's suggestions and case studies for schoolteachers. The suggestions include projects linking home and school communities, involving parents and NGOs in environmental education (QCA 2003). A second example of environmental education practice offering outdoor experiences *and* developing helpful interpersonal relationships is the educational work of some NGOs. A recent example is exemplified in the work of the non-profit organization, The Environmental Alliance for Senior Involvement (EASI 2003). The United States program has links with the US Environmental Protection Agency as well as organizations in the United Kingdom and the Netherlands. The EASI volunteers of over 55 years old are involved in over 350 programs in the United States, engaging senior citizens in hands-on environmental care activities. Almost all programs include an intergenerational element wherein young members are taught by "elders" or "environmental stewards" of the community. The activities of this program reflect research evidence concerning the positive effects of intergenerational learning. The EASI program also illustrates that learning can flow from elder to youth as well as from young student to adult, as suggested by Ballantyne *et al.* (1998). Horwitz (1996) and Chawla (1999) provide other examples from the United States and Norway respectively.

CONCLUSION

Assessing the relationship between recent environmental education research and educational practice is a far from easy or straightforward task. First, the translation of research evidence directly into practice is complex. Additionally, as pointed out by Rickinson (2001), weaknesses attributable to the past decade's "action" phase of research encumber the direct application of findings to practice. Such weaknesses include: (a) a lack of longitudinal work and the need for follow-up work after evidence-gathering; (b) the need for individual studies to become more comparable in methodology and also the need to expand the range of qualitative methodologies employed; (c) a current over-reliance upon interpretivist and/or constructivist approaches in qualitative research; (d) a shortage of research into the sources of young peoples' learning processes, and behaviors alongside the study of their existing experiences, knowledge, and attitudes and behaviors relating to environmental learning; and (e) insufficient dialog between environmental education researchers and users. These weaknesses can be transferred into opportunities and suggestions for future research, but for practitioners of environmental education they represent significant obstacles to be overcome.

Additionally, practitioners are confronted with the tasks of unraveling necessary, but nevertheless confusing, arguments about the concept, use, and status of education for sustainability (Jickling 1992; Martin 1996; Maser 1996; Fien and Tilbury 2002; McKeown and Hopkins 2003; Sterling 2003). Contentions within research dialog make it no easier for practitioners, especially within the more constricted settings of formal education, to make well-informed decisions about how to present environmental education opportunities in practice.

However, in a far more positive vein, this chapter has provided examples of some of the more visible applications of research to proposed and actual practice that have emerged over the past few years. In 1999 Palmer commented on the urgent need to increase our empirical database through programs of high quality research in environmental education and then to find ways of enabling this research base to improve the promise and potential of educational practice. These challenges remain high on the agenda of environmental educators around the globe. Yet without doubt progress is being made.

Researchers are increasingly engaged in grounding methodologies and connecting research with practice by using a broader base of research methodologies and embracing more diverse aspects of environmental understandings than ever before. Social contexts are at the forefront of a good number of reported investigations and the increasing number of large-scale funded international projects will inevitably have impact upon future research, policy, and practice both within and across national boundaries. Perhaps the field of environmental education research will have "come of age" when firm evidence suggests that it provides the context for effective planning, implementation, and on-going evaluation of programs of environmental education within which formal and informal elements are mutually supportive.

REFERENCES

Arbuthnot, J. (1977). The roles of attitudinal and personality variables in the prediction of environmental behaviour and knowledge. *Environment and Behaviour*, **9**(2), 217–232.

Arcury, T. A. and Christianson, E. H. (1993). Rural-urban differences in environmental knowledge and actions. *Journal of Environmental Education*, **25**(1), 217–232.

Ballantyne, R., Connell, S., and Fien, J. (1998). Students as catalysts of environmental change: a framework for researching intergenerational influence through environmental education. *Environmental Education Research*, **4**(3), 285–298.

Bloomfield, P. (1998). Raising awareness of local Agenda 21. In *Primary Sources. Research Findings in Primary Geography*, ed. S. Scoffham. Sheffield, UK: The Geographical Association, pp. 34–35.

Bogner, F. X. (1998). The influence of short-term outdoor ecology education on long-term variables of environmental perspective. *Journal of Environmental Education*, **26**(1), 17–29.

Bonnett, M. and Williams, J. (1998). Environmental education and primary children's attitudes towards nature and the environment. *Cambridge Journal of Education*, **28**(2), 159–174.

Caro, T., Pelkey, N., and Grigione, M. (1994). Effects of conservation biology education on attitudes towards nature. *Conservation Biology*, **8**(3), 846–852.

Chawla, L. (1998). Significant life experiences revisited: a review of research on sources of environmental sensitivity. *Journal of Environmental Education*, **29**(3), 11–21.

(1999). Life paths into environmental action. *Journal of Environmental Education*, **31**(1), 15–26.

Chipeniuk, R. (1998). Lay theories of spring: displacement of common-sense understandings of nature by "expert" ideas. *International Research in Geographical and Environmental Education*, **7**(1), 14–25.

Connell, S., Fien, J., Sykes, H., and Yenken, D. (1998). Young people and the environment in Australia: beliefs, knowledge, commitment and educational implications. *Australian Journal of Environmental Education*, **12**, 19–26.

Csobod, E. (2002). Hungary: an education initiative for a sustainable future. In *Education and Sustainability: Responding to the Global Challenge*, ed. D. Tilbury, B. Stevenson, J. Fien, and D. Schreuder, pp. 99-107. Gland, Switzerland: IUCN, pp. 99–107.

Dettman-Easler, D. and Pease, J. (1999). Evaluating the effectiveness of residential environmental education programs in fostering positive attitudes toward wildlife. *Journal of Environmental Education*, **31**(1), 33–39.

EASI (2003). *The Environmental Alliance for Senior Involvement.* [Online] URL: http://www.easi.org/(retrieved March 2003).

Elliot, J. (1991). *Developing Community-focused Environmental Education Through Action-research.* Mimeograph. Norwich: Centre for Applied Research in Education, School of Education, University of East Anglia.

Emmons, K. M. (1997). Perceptions of the environment while exploring the outdoors: a case study in Belize. *Environmental Education Research*, **3**(3), 327–344.

Esland, G. (1971). Teaching and learning as the organization of knowledge. In *Knowledge and Control: New Directions for the Sociology of Education*, ed. M. F. D. Young. London: Collier-Macmillan.

Evans, S. M., Gill, M. E., and Marchant, J. (1996). Schoolchildren as educators: the indirect influence of environmental education in schools on parents' attitudes towards the environment. *Journal of Biological Education*, **30**(4), 243–249.

Fensham, P. (1978). Stockholm to Tbilisi: the evolution of environmental education. *Prospects*, **8**(4), 446–455.

Fien, J. (1992). *Education for the Environment: A Critical Ethnography.* Brisbane, Australia: University of Queensland.

Fien, J. and Hillcoat, J. (1996). The critical tradition in research in geographical and environmental education research. In *Understanding Geographical and Environmental Education*, ed. M. Williams, London: Cassell, pp. 26–40.

Fien, J. and Tilbury, D. (2002). The global challenge of sustainability. In *Education and Sustainability: Responding to the Global Challenge*, ed. D. Tilbury, B. Stevenson, J. Fien, and D. Schreuder. Gland, Switzerland: IUCN, pp. 1–13.

Gough, N. (1999). Rethinking the subject: (de) constructing human agency in environmental education research. *Environmental Education Research*, **5**(1), 35–48.

Gunderson, K. I. (1989). *The state of environmental education in Montana public schools*, K-6. Unpublished Masters Thesis, University of Montana, Missoula.

Hart, R. (1997). *Children's Particpation: The Theory and Practice of Involving Young Citizens in Community Development and Earth Care*. London: Earthscan.

Hines, J., Hungerford, H., and Tomera, A. (1986/7). Analysis and synthesis of research on responsible environmental behaviour: a meta-analysis. *Journal of Environmental Education*, **18**(2), 1–8.

Horwitz, W. (1996). Developmental origins of environmental ethics: the life experiences of activists. *Ethics and Behaviour*, **1**(6), 29–54.

Huckle, J. (1993). Environmental education and sustainability: a view from critical theory. In *Environmental Education: A Pathway to Sustainability*, ed. J. Fien. Geelong, Australia: Deakin University Press, pp. 43–68.

Hungerford, H. and Volk, T. (1990). Changing learner behaviour through environmental education. *Journal of Environmental Education*, **21**(3), 8–21.

Hungerford, H., Peyton, R., and Wilkie, R. (1980). Goals for curriculum development in environmental education. *Journal of Environmental Education*, **2**(3), 42–47.

(1983). Editorial. Yes EE does have a definition and structure. *Journal of Environmental Education*, **14**(3), 1–2.

Iozzi, L. A. (1981). *Research in Environmental Education 1971–1980*. (ED214762). Columbus, OH: ERIC Clearinghouse for Science, Mathematics and Environmental Education.

IUCN. (1980). *World Conservation Strategy*. Gland, Switzerland: IUCN.

IUCN, UNEP, WWF (1991). *Caring for the Earth: A Strategy for Sustainable Living*. Gland, Switzerland: IUCN.

Jickling, B. (1992). Why I don't want my children to be educated for sustainable development. *Journal of Environmental Education*, **23**(4), 5–8.

Juniper, A. (1989). Marketing nature conservation. *Ecos. A Review of Conservation*, **10**(1).

Keliher, V. (1997). Children's perceptions of nature. *International Research in Geographical and Environmental Education*, **6**(3), 244–246.

Kidd, A. H. and Kidd, R. M. (1997). Characteristics and motives of adolescent volunteers in wildlife education. *Psychological Reports*, **80**, 747–753.

Leeming, F. C., Porter, B. E., Dwyer, W. O., Cobern, M. K., and Oliver, D. P. (1997). Effects of participation in class activities on children's environmental attitudes and knowledge. *Journal of Environmental Education*, **28**(2), 33–42.

Lucas, A. M. (1979). *Environment and Environmental Education: Conceptual Issues and Curriculum Interpretations*. Kew, Australia: Australia International Press and Publications.

Marcinkowski, T. (1990). A contextual review of the "quantitative paradigm" in EE research. Paper presented at a symposium entitled Contesting Paradigms in Environmental Education Research, *Annual Conference of the North American Association for Environmental Education*, San Antonio, Texas.

Martin, P. (1996). A WWF view of education and the roles of NGOs. In *Education for Sustainability*, ed. J. Huckle and S. Sterling, London: Earthscan Publications, pp. 40–51.

Maser, C. (1996). *Resolving Environmental Conflict: Towards Sustainable Community Development*. Defray Beach, FL: St. Lucie Press.

McKeown, R. and Hopkins, C. (2003). EE≠ESD: diffusing the worry. *Environmental Education Research*, **9**(1), 117–128.

Morris, M. and Schagen, I. (1996). *Green Attitudes or Learned Responses?* Slough, UK NFER.

Palmer, J. A. (1993). Development of concern for the environment and formative experiences of educators. *Journal of Environmental Education*, **24**(3), 26–30.

(1995). Environmental thinking in the early years: understanding and misunderstanding of concepts related to waste management. *Environmental Education Research*, **1**(1), 35–45.

(1998a). *Environmental Education in the 21st Century: Theory, Practice, Progress and Promise*. London: Routledge.

(1998b). Spiritual ideas, econcerns and educational. In *Spirit of the Environment: Religion, Value and Environmental Concern*, ed. D. E. Cooper and J. A. Palmer. London Routledge, pp. 146–147.

(1999). Research matters: a call for the application of empirical evidence to the task of improving the quality and impact of environmental education. *Cambridge Journal of Education*, **29**(3), 379–395.

Palmer, J. A. and Lancaster, J. C. (1999). *Review of the Educational Activities of the Wildlife Trusts*, UK. Unpublished Final Report, The University of Durham. September.

Palmer, J. A. and Neal, P. (1994). *The Handbook of Environmental Education*. London: Routledge.

Palmer, J. A., and Suggate, J. (1996). Influences and experiences affecting the pro-environmental behaviour of educators. *Environmental Education Research*, **2**(1), 109–121.

Palmer, J. A., Suggate, J., and Matthews, J. (1996). Environmental cognition: early ideas and misconceptions at the ages of four and six. *Environmental Education Research*, **2**(3), 301–329.

Palmer, J. A., Suggate, J., Bajd, B., Hart, P., Ho, R. K. P., Ofwono-Orecho, J. K. W., Peries, M., Robottom, I., Tsaliki, E., and Van Staden, C. (1998). An overview of significant influences and formative experiences on the development of adults' environmental awareness in nine countries. *Environmental Education Research*, **4**(4), 445–464.

Palmer, J. A., Suggate, J., Bajd, B., Tsaliki, E., Mati, D., Paraskovopoulos, S., Razpet, N., and Skribe, D. (1999). Emerging knowledge of distant environments: an international study of four and six year olds in England, Slovenia and Greece. *Journal of the European Early Childhood Education Research Association*, **7**(2), 17–29.

Peterson, N. and Hungerford, H. (1981). Developmental variables affecting environmental sensitivity in professional environmental educators. In *Current Issues in Environmental Education and Environmental Studies*, ed. A. B. Stacks, L. A. Iozzi, J. M. Scjhultz, and R. Walks Columbus, OH: ERIC, pp. 111–113.

Qualifications and Curriculum Authority (QCA) (2003). *Education for Sustainable Development Case Studies*. [Online] URL: http://www.nc.uk.net/esd/teaching/case_studies/case_study_g.htm (retrieved April 2003).

Rickinson, M. (2001). Learners and learning in environmental education: a critical review of the evidence. *Environmental Education Research*, **7**(3), 207–320.

Robottom, I. (1982). What is environmental education as education about the environment. Paper presented at the *Second National Conference of the Australian Association for Environmental Education*. Brisbane.

(1983). *The Environmental Education Project Evaluation Report*. Curriculum Development Centre, Canberra/Deakin University, Victoria.

Robottom, I. and Hart, P. (1993). *Research in Environmental Education: Engaging the Debate*. Geelong, Australia: Deakin University.

Rovira, M. (2000). Evaluating environmental education programs: Some issues and problems. *Environmental Education Research*, **6**(2), 143–156.

Schreuder, D., Reddy, C., and Le Grange, L. (2002). Environmental education as a process of change and reconstruction: the science and sustainability project.In *Education and Sustainability: Responding to the Global Challenge*, ed. D. Tilbury, B. Stevenson, J. Fien, and D. Schreuder. Gland, Switzerland: IUCN, pp. 133–139.

Scott, D. and Willits, F. K. (1994). Environmental attitudes and behaviour: a Pennsylvanian survey. *Environment and Behaviour*, **26**(2), 239–60.

Stapp, W., Caduto, M., Mann, L. and Nowail, P. (1980). Analysis of pre-service environmental education of teachers in Europe and an instructional model for furthering this education. *Journal of Environmental Education*, **12**(2), 3–10.

Sterling, S. (2003). *WWF Learning News – Making It or Faking It?* [Online] URL: http://www.wwflearning.co.uk/news/features_0000000350.asp (retrieved March 2003).

Stevenson, R. B. (1987). Schooling and environmental education: contradictions in purpose and practice. In *Environmental Education: Practice and Possibility*, ed. I. Robottom. Geelong, Australia: Deakin University Press, pp. 69–82.

Sward, L. (1996). Experiential variables affecting the environmental sensitivity of El Salvadoran environmental professionals. Paper Presented at *the Annual Conference of the North American Association of Environmental Education*. November 1996, Burlington, California.

Tanner, T. (1980). Significant life experiences. *Journal of Environmental Education*, **1**(4), 20–24.

UNCED (1992). *Agenda 21*. United Nations Conference on Environment and Development (The Earth Summit). New York, NY: United Nations Publications.

UNESCO (1975). *The International Workshop on Environmental Education Final Report*. Belgrade, Yugoslavia. Paris: UNESCO/UNEP.

(1977). *First Intergovernmental Conference on Environmental Education Final Report*. Tbilisi, USSR. Paris: UNESCO.

Uzzell, D., Davallon, J., Fontes, P. J., Gottesdiener, H., Jensen, B. B., Koefoed, J., Uhrenholdt, G., and Vognsen, C. (1994). *Children as Catalysts of Environmental Change: Report of an Investigation on Environmental Education*. Final Report. Brussels: European Commission.

Uzzell, D., Rutland, A., and Whistance, D. (1995). Questioning values in environmental education. In *Values and the Environment: A Social Science Perspective*, ed. Y. Guerrier, N. Alexander, J. Chase. and M. O'Brien. Chichester:, UK: John Wiley, pp. 171–182.

Volk, T. L., Hungerford, H. R., and Tomera, A. N. (1984). A national survey of curriculum needs as perceived by professional environmental educators. *Journal of Environmental Education*, **16**(1), 10–19.

Votaw, T. (1983). *Antecedents to environmental attitudes: a survey of seasonal naturalists at Denali National Park and Preserve*. Las Cruces, NM: Unpublished research Paper.

JOHN F. DISINGER

7

The purposes of environmental education: perspectives of teachers, governmental agencies, NGOs, professional societies, and advocacy groups

INTRODUCTION

Educators should be well positioned to decide for themselves what they will present in their classes and how they will do it, given their combination of professional training and academic preparation in the subject areas in which they teach. But they are subject to continuing influences from a number of external sources, some obligatory and obvious, others less obligatory and less obvious. This chapter touches on the former, but focuses on the latter.

At the business meeting of the 1978 conference of the National Association for Environmental Education at Estes Park, Colorado, an Association member initiated a politically oriented, pro-environmental resolution on a controversial environmental issue, which after a heated debate was voted down. In a plenary session at a 1983 environmental education conference coordinated by the (National) Alliance for Environmental Education at the University of Vermont, a similarly oriented proposal advanced by a conference participant also was defeated. In neither situation was the individual who advanced the proposal a professional educator, but both were regularly involved in the activities of environmental education organizations, and both also were active in environmental advocacy organizations.

What was there about "environmental education" that attracted the involvement of people who were not professional educators? Why did they advance *politically oriented initiatives* at meetings designed, or at least advertised, to address *educational* issues? How clear were the distinctions between environmental advocacy and environmental education among conference attendees? Why were these proposals defeated?

To what extent should the reemergence of such situations be anticipated, and (if they do occur) what should be the responses of professional educators?

At least in the United States, environmental advocates have looked at environmental education as a potentially useful vehicle for advancing their proenvironment positions. In neither situation discussed above did the arguments leading to proposal rejection deal with its scientific-economic-political-environmental merits. However, the telling arguments were that the goals of the sponsoring organizations and the purposes of the conferences themselves were for education, not for environmentalism. In both cases, those opposing the proposals identified the absence of professional-level expertise in the scientific-economic-political aspects of interactions between humans and environment on the part of those attending the conference as a critical deficiency. These arguments raised more questions. Should educational organizations themselves ever claim the technical expertise to take formal positions on matters other than those dealing directly with education? Or should individual educators themselves do so, by virtue of their status as professionals in education?

HISTORICAL BACKGROUND

If one reviews the history of environmental education back to its identifiable predecessors, the nature study and outdoor education movements of the late nineteenth and early twentieth centuries, one finds little evidence that public positions on non-pedagogic issues, environmental or otherwise, were endorsed by educators until quite recently. Proponents of nature study and outdoor education have historically believed that it is educationally appropriate to teach children about the natural world, and pedagogically sound to take them outside the school building to do it, but historically they did not publicly take positions on environment-related or other issues of public concern, and non-educators did not suggest that they do so. Until recently, little consideration was given by professional educators to the idea of speaking out on controversial matters, in the classroom or elsewhere. Likewise, people who were not educators were unlikely to become directly involved in education to the extent that they sought to influence the curriculum in any manner approaching the initiation or support of political activism.

This does not mean that schools and educators have not historically endorsed value-laden positions. Through the nineteenth century and well into the twentieth, values commonly held by American society and explicitly supported in school curricula included the enthusiastic, essentially unquestioned acceptance of the idea that the natural environment appropriately served as a reservoir of readily available resources whose primary function was to serve as a God-given cornucopia for the development of a materially better life for all. Nor was environmental quality seen as a matter of urgency by either society or educators; rather, it was generally agreed that (for example) "the solution to pollution is dilution," so that releasing waste materials into the environment was typically seen as an effective, essentially harmless way of disposing of waste materials.

Socially accepted values have for centuries routinely been infused in school curricula and programs, but since neither resource depletion nor environmental quality was historically perceived as of significant concern to American society, neither found a place in the priorities of either society or schools until recently. Beginning in the latter half of the nineteenth century, environmental quality arguments were advanced in the public arena by nature-focused writers such as Henry David Thoreau, John Muir, and John Burroughs, but if their works were mentioned at all in schools it was because their authors were perceived as gifted but philosophically obtuse writers expressing quaint worldviews that marked them as eccentrics, out of tune with the mainstream thinking of their days. In schools as in society, any consideration of their works was in the "impractical" framework of poetic-quality literature, not the "more practical" concerns associated with territorial expansion and economic and technological development.

Pleas for the wise-use management of natural resources began to appear in American society when the first American conservation movement emerged in the late 1800s, nurtured by George Perkins Marsh, John Wesley Powell, Gifford Pinchot, and others. Utilitarian resource management attracted significant public and political attention for a time but, because it challenged the dominant value system of the day it received little consideration in schools: "To a large degree the educational philosophy of the period ruled against the consideration of current problems in the schools. Hence, little attention was paid in the schools to what would [eventually] be termed conservation education . . ." (Swan 1975: 7–8).

Though education about utilitarian conservation was unable to establish a foothold in schools in the early years of the twentieth century, nature study did so. Its early success may be attributed at least in part to the fact that it was not seen as controversial, so it was not a problem in that respect. The American Nature Study Society

(ANSS), formed in 1908, attracted persons (authors, college professors, elementary and high school teachers, hobbyists, and others) "interested in developing a strong program of elementary science in the schools with emphasis on nature study rather than watered-down technical science" (Russell 1978: 45). Through ANSS and other nature-related organizations, conferences, publications, and networks of committed individuals, only some of whom are professional educators, nature study has maintained a consistent, positive, low-key presence in schooling for more than a century, from the late 1800s to the present.

EDUCATION FOR SOCIAL RESPONSIBILITY?

During the Great Depression of the 1930s, perceptions of how schools should deal with social problems began to change. In 1935, as renewed public and political interest in the utilitarian conservation of natural resources gathered steam, the National Education Association's Educational Policies Commission stated that "the schools may well assume considerable responsibility for checking the ravages upon the [natural] heritage of the nation made by ignorance, indifference, carelessness, and unbridled selfishness" (Funderburk 1948: 28). This was a clear indication that leaders within the educational establishment felt that schools should assume a role in promoting social change, in support of an emerging shift in social values. Thus, a "Dust Bowl mindset" emerged, giving rise to conservation education. Its purpose was to awaken the public, including school children, to the importance of utilitarian resource management and the necessity of preventing human-induced natural disasters.

During the 1930s, recently created or newly reinvigorated federal and state natural resource management agencies recognized that a public educated to understand the values of resource conservation could be instrumental in accomplishing both conservation goals and agency missions, and realized also that the younger members of that public ("the voters of tomorrow") were in schools. It was clear that the schools could be enlisted to promote environment-related agendas, *and that educators were willing to join up*. Thus governmental agencies began to promote their government-mandated management agendas in the schools. Citizen organizations also welcomed the opportunity to use the schools as venues for promotion of their conservation-preservation priorities. Quite often, they were able to initiate and support state legislation requiring schools to incorporate conservation education in their curricula. Beginning in the 1930s, "leadership for conservation education came from many sources and from many organizations, especially governmental and private agencies and associations outside of schools, such as the soil and water conservation districts, soil conservation service, wildlife federations, and sportsmen's clubs" (Swan 1975: 9).

The Conservation Education Association (CEA) was founded in 1953 to promote conservation education "at all levels, encompassing all aspects of the natural and man-made world upon which people rely for the development and maintenance of a desirable social, economic, scientific, cultural and political climate" (Myshak 1978: 60). Both its membership and its leadership consisted of a mixture of professional educators, professional natural resource managers, public information specialists employed by governmental agencies and resource-associated industries, and representatives of non-governmental organizations (NGOs). Within CEA these groups worked together to promote education about balanced interrelationships between humans and the natural environment. Over the years representatives of federal and state agencies became increasingly influential in CEA's activities, and through CEA had significant influence on conservation-related programming in schools.

Interrelationships among CEA members were generally characterized by low-key, collegial give-and-take. Resource managers and educators, mostly classroom teachers, shared complementary professional knowledge of conservation-related content and effective educational methodologies, as well as an often passionate commitment to teaching children about the environment and environmental responsibility.

In the meantime, the progressive education movement sought to reform traditional teaching–learning methodologies, which historically were based on lecture-memorization-drill-recall-recitation approaches, by promoting direct contact with things, places, and people, the purpose being to help each individual student to develop physically, intellectually, socially, and emotionally. Both nature study and outdoor education were obvious "fits" with the philosophies and methodologies of progressive education because they focused on direct experience in the one case and non-traditional approaches to learning in the other. When nature study took place outside the classroom, the congruencies were even greater.

Though progressive education as originally conceptualized focused on holistic learning for the development of the individual child, some of its followers "chose to direct the movement toward radical social reform, using the school system as a tool" (Roth 1978: 17). The resulting backlash initiated a controversy that continues today – to what extent should agendas involving society-wide political, economic, and social change be considered in the classroom? Should they be promoted, endorsed, ignored, or rejected? Is the most important purpose of the school to perpetuate the culture in which it exists or to re-form that culture, or should it attempt to do both? If both, what balance should be sought? Though progressive education no longer exists as an identifiable and widespread movement, those questions persist.

AN EMERGING AGENDA

The American educational system itself became a target of strong criticism when in 1956 the Soviet Union launched its first space satellite. The argument essentially was that American science and technology would have produced innovative space-age technologies sooner than the Soviets if its scientists had been better educated, and they would have been better educated if the schools had taught them more rigorous science and mathematics. This led to a reduction in emphasis on, if not the elimination of "less quantitative" curriculum components to make available more classroom time and curricular space for more advanced, more highly quantitative studies. Under these circumstances priorities for the arts and humanities diminished, progressive education was largely forgotten, and conservation education and nature study were seen by many as largely irrelevant.

Thus a series of high profile, highly rigorous science curriculum projects was initiated to replace more descriptive, more qualitative science courses with conceptually based, quantitatively mature courses. There were commonalities among the new courses. Generally they were bankrolled by the National Science Foundation (NSF) and were headed intellectually and administratively by professional scientists, experts in their fields but not necessarily in education. The goal of these courses was to prepare learners at progressively more advanced levels as they moved through their K-12 years and ultimately into post-secondary educational institutions. PSSC Physics for high school seniors appeared first, CHEM Study chemistry for juniors followed, and so on, with programs and materials for the elementary school (for example, Science: A Process Approach) emerging at later dates.

In point of fact, the educational system was a convenient scapegoat for what was a reorientation of public and political priorities. Education administrators, curriculum specialists, and teachers were "punished" for their perceived failures by being treated as if they themselves were impediments to the curriculum redesign process. At first, the developers of these projects attempted to work directly with classroom teachers in individual schools, bypassing as much as they could the professional education hierarchies of colleges, universities, state education agencies, and local administrators. The scientists who provided project leadership seized, carpe diem, the unanticipated but warmly welcomed opportunity to improve student preparation for post-secondary studies in their own specialized fields. They were both energetic and creative, and there is no doubt that they were successful in terms of increasing the rigor of K-12 science courses. Though parts of what they initiated continue to exist in today's science curricula, much of what was accomplished in the short term eventually was dissipated because the programs were not effectively institutionalized. Because the initiators were "outsiders" who made few attempts to interface with leaders in the professional education establishment, they were unable to sustain their programs when NSF funding was no longer available.

Educational interest in the study of the environment was curtailed, but not obliterated, by this onslaught of academic rigor during the 1950s. During this time period, nature study, outdoor education, and conservation education maintained modest niches, particularly in elementary schools, materially aided in doing so by the efforts of organizations like the American Nature Study Society and the Conservation Education Association.

ENVIRONMENTALISM

The publication in 1962 of Rachel Carson's *Silent Spring* initiated a substantially more energetic wave of public concern about environmental matters. This caused a pronounced shift from the utilitarian concerns that had dominated the conservation movements of the 1890s and the 1930s to a substantially greater emphasis on environmental quality and human health (Hays 1985). As might be expected, the emerging movement spawned an interest in education relating to the environment, but no early consensus on how to approach it.

Still, environment-related concerns impacted the educational establishment at all levels. Included was a resurgence of attention in schools to nature and utilitarian conservation, but more striking was the development of aggressive attention to contemporary environmental problems and how to deal with them. A widespread feeling emerged that the educational system should be actively involved in society's quest for the improvement of environmental quality in much the same way that schools had been expected to play a role in the conservation of natural resources in the 1930s. This feeling was shared by many scientists, educators, and environmentalists, as well as by public officials and political leaders.

Identifying a viable approach for schools to deal with environmental matters was elusive, as was devising methodologies to do so. Lucas (1972) characterized three prominent approaches, investigating education *about* the environment, education *in* the environment, and education *for* the environment, with a number of permutations and combinations of those emphases. Conclusions of his study suggested that education *for* the environment deserved a significant place in formal education, but faced difficult hurdles, including institutional inertia and curricular territorialism, before it could enter the educational mainstream.

Earlier, Stapp and his students had defined environmental education to include a "motivation to act" dimension (Stapp *et al.* 1969), and since that time the "action" component - education *for* the environment - has been verbalized as an article of faith by many of those who consider themselves environmental educators. Their assumption is that if students are properly taught about the environment, they will subsequently behave in an environmentally responsible manner and engage in proenvironmental actions, at least in terms of their personal activities. Similar assumptions underlie legislation enacted by the US Congress, including the parts of the National Environmental Policy Act of 1969 that called for "enrich[ing] the understanding of the ecological systems and natural resources important to the Nation" (US Public Law 91-190, 1970: 1), and all of the Environmental Quality Education Act of 1970, which described environmental education as "the educational process dealing with man's relationship with his natural and manmade surroundings, and includes the relation of population, conservation, transportation, technology, and urban and regional planning to the total human environment" (US Public Law 91-516, 1970: 1).

Though both of these pieces of legislation stressed a need for education as a social institution to take a leading role in developing understanding of the dimensions of environmental problems, neither called on educators to take an active role in solving them. Rather, both stressed the need for enhanced ecological-environmental knowledge, and implied that the task of education was to see that such knowledge was made available; that is, schools must provide education *about* the environment and its associated problems. The implicit, and somewhat naïve, assumption was that if students learn about the environment and problems related to it they will automatically behave in such a manner that the problems will be resolved, though it has long been known that a constellation of additional factors are involved. These are addressed later in this chapter.

MEETING THE NEED?

In 1970, the US Office of Education created an Office of Environmental Education (OEE) to implement the provisions of the Environmental Education Act. OEE administered a small grants operation, the idea being that it would furnish seed grants to innovative educational projects. But it was (and still is) difficult to establish sustainable, innovative programs with limited one-time funds. Further, grants often were awarded to enthusiastic but impractical groups attempting to influence education from the outside – for example, to environmental advocacy groups who had little understanding of the purpose, practice, and practicalities of education. As is often the case with legislation passed with great fanfare and stirring rhetoric, the Environmental Education Act of 1970 was unaccompanied by funding sufficient even to approach the accomplishment of its task, and was operational for only a few relatively ineffective years.

The more recent US Environmental Education Act (P.L. 101-619, 1990) also established an Office of Environmental Education (also OEE), but housed it in the US Environmental Protection Agency (EPA) rather than the Department of Education. In EPA, the current OEE is viewed at least in part as a mechanism through which the host agency's environmental protection mission may be advanced, i.e., education as a tool, rather than as an end in itself. That said, the current OEE maintains a high priority on working closely with practitioners in the field. Given the magnitude of its task, OEE is again underfunded. Nonetheless, it has established a national advisory board composed of professional educators, maintains a small grants program, sponsors an awards program for individuals, and works with a National Environmental Education and Training Foundation (NEETF) to channel private funds into environmental education activities deemed consistent with EPA's goals. OEE has been in danger of termination on several occasions, probably because its activities are sometimes seen as more political and activist than educational.

International intergovernmental bodies also have sought to enlist educators in education *for* the environment. For example, Recommendation 96 of the 1972 Stockholm Conference on the Human Environment, an international forum initiated by UNESCO, called for the development of environmental education as a critical element in an all-out attack on the world's environmental crisis, to "creat[e] citizenries not merely aware of the crisis of overpopulation, mismanagement of natural resources, pollution, and degradation of the quality of human life, but also able to focus intelligently on the means of coping with them" (UNESCO 1972: 10–11). Similarly, the Tbilisi Declaration, adopted at a 1977 UNESCO/UNEP intergovernmental conference on environmental education in Georgia, USSR, stated that:

> A basic aim of environmental education is to succeed in making individuals and communities understand the complex natures of the natural and the built environments, resulting from the interaction of their biological, physical, social, economic and cultural aspects, and acquire the knowledge, values, attitudes and practical skills to participate in a responsible and effective way in anticipating and solving environmental problems, and in the management of the quality of the environment. (UNESCO 1978: 1)

ORGANIZATIONAL RESPONSES: GOVERNMENTAL AGENCIES, NGOS, PROFESSIONAL SOCIETIES

In the governmental sector, a Subcommittee on Environmental Education (SEE) of the Federal Interagency Committee on Education (FICE) was initiated in 1975. Through FICE/SEE, representatives of 20-plus scientific-technical-managerial agencies of the US federal government endeavored:

> To improve cooperation, coordination, and avenues of information exchange among Federal agencies in meeting the nation's environmental education needs and goals and to establish linkages and promote a ready exchange of concerns among the various FICE members and nonfederal groups in order to foster and encourage the formation of a national strategy on environmental education, and to assist and stimulate bilateral and multilateral environmental education efforts among nations. (Jeske 1978: 162)

One of FICE/SEE's projects was the development of "a coherent series of concepts that citizens must understand in order to think intelligently about and act responsibly toward their environment" (Jeske 1978: 159). The resulting document, *Fundamentals of Environmental Education* (FICE/SEE 1976), called on the formal education system to play a major role in education for environmental quality, as conceptualized by representatives of the participating federal agencies. FICE/SEE also coordinated US preparatory efforts for the Tbilisi Intergovernmental Conference on

Environmental Education in 1997, by involving some 70 environmental education leaders from federal and state governments, industry, academia, and non-governmental organizations in advance planning. FICE/SEE was a "soft-sell" organization that sought to further the environment-related goals of its member agencies by explaining their missions carefully and capitalizing on the interests they shared with other agencies, NGOs, and the environmental education community, all the while being careful not to compromise their own missions and goals.

In the private sector, CEA was instrumental (in 1973) in the conceptualization and initiation of the Alliance for Environmental Education (AEE), a coalition of NGOs that sought to promote the environment-related agendas of its members through involvement in schooling, primarily by offering printed and audio-visual materials, advisory services, and program participation (McSwain 1978). Of the 32 member organizations self-reported in a 1979 compendium of the activities of AEE affiliates (Disinger 1978), 12 were closely associated with professional education, nine were traditional conservation-focused non-governmental organizations, three were industry public relations/information groups, two were public information/education groups, two were advocacy groups with environment-related concerns, two were large membership youth organizations, one was a professional association with expertise in the built environment, and one was the conservation department of a major labor union.

AEE's activities were generally conducted in a collegial manner and were well received by those who identified themselves as environmental educators, though the "umbrella" nature of the organization meant that those participating were dealing with leaders of member groups more frequently than with classroom practitioners or other professional educators in general. Beyond that, the Alliance was neither well known nor particularly influential within the larger professional education community, as may be said for most organizations that have attempted to assume leadership in environmental education. This may be because environmental education is effectively interdisciplinary, while the professional educational system operates largely within a disciplinary framework.

In addition to serving as a communications coordination body among member organizations and environmental educators, AEE coordinated several national (US) conferences between the middle 1970s and early 1990s, but eventually dissolved when it was unable to generate funding sufficient to sustain an increasingly aggressive agenda that sought to change it into a pro-active NGO with education-oriented programs and activities. The upper management levels of AEE member organizations apparently viewed their involvement in education as a public relations function and did not accord sufficiently high priority to cooperative involvement in educational activities to support them at substantial levels, though many supported specific educational activities within the scope of their own narrower interests. A noteworthy example is *Project Learning Tree*, a teaching guide funded by the American Forest Institute (AFI) and developed by the Western Regional Education Council (WREEC), with substantial AFI input (WREEC 1975).

In 1971, a group of two-year college administrators decided to promote environment-related instructional materials that they had designed for use at the community college level by organizing an association specifically for the purposes of promotion of those materials. However, after they picked "National Association for Environmental Education" (NAEE) as the organization's name they found themselves inundated with educators representing all educational levels, along with non-educators with proenvironmental interests. The unanticipated influx of individuals with different agendas quickly broadened the association's purview to reflect the multiple interests represented – environmentalists, employees of governmental agencies, business and industry representatives, communications, public information, and public relations specialists, outdoor educators, nature interpreters, conservation educators, teacher educators, curriculum specialists, and elementary, secondary, and post-secondary teachers from a variety of disciplines (Disinger 2001). NAEE's resulting lack of focus, and its subsequent and ongoing efforts to be inclusive of all who expressed an interest in environmental education, was typical of the state of the field at that time, and since that time.

Over the three decades of its existence, the organization has changed its name and geographic scope once (it is now the North American Association for Environmental Education, NAAEE) and its leading segments several times, from community colleges to non-formal practitioners to classroom practitioners to college/university environmental scientists to environmental communicators to curriculum specialists to teacher educators, not necessarily in that order, and not necessarily one group at a time. Being all things to all people was not then and is not now easy, but that seems to be NAAEE's *modus operandi*, and indeed that of most groups that have tried to assume leadership in the field. In the recent past, NAAEE has demonstrated increasingly creative leadership as it has made significant strides in professional productivity, the most obvious being the development of guidelines for teaching materials (NAAEE 1996), the publication of objective reviews of resources for instruction in environmental education

(NAAEE 1997, 1998), and an increasingly comprehensive practitioner-friendly website (www.naaee.org). Other significant accomplishments include the demonstration of rigorous professionalism in its interactions with both supporters and critics of environmental education.

BACKLASH, QUESTIONS OF SUBSTANCE, AND RESPONSE

The 1980s saw the emergence of an anti-environment movement primarily based on conservative political ideologies. Public opinion surveys showed that the general public continued to express support for idealistic proenvironmental agendas (Dunlap 1987a), but also that it placed a higher priority on "practical matters" (e.g., economics and national security) so that the environment became, in fact, a "second-order issue" in the eyes of the public (Dunlap 1987b).

Spokespersons for the anti-environmental movement took the position that many teachers were promoting environmental advocacy in their classrooms, according to an analysis by Coyle (1996). The extent to which teachers actually teach (or "preach") environmentalism in their classrooms has not been statistically documented, though it was vividly demonstrated anecdotally (Sanera and Shaw 1996). Both the Coyle and Sanera–Shaw arguments deserve careful attention, because they emphasize vividly contrasting perspectives on the meager evidence available. Sanera, of the Claremont Institute, and Shaw, of the Political Economy Research Center, claimed that some percentage (specifics lacking) of what passes for "environmental education" is inaccurate and emotionally charged, primarily because it is derived from one-sided, proenvironmental sources. But NEETF's Coyle responded that the anti-environment groups offered a similarly inaccurate view, primarily because it is derived from one-sided, anti-environmental sources.

Another recurring criticism of environmental education is that it lacks substance. That is, when the environment is the topic, academic rigor is often replaced with simplistic, sometimes erroneous, often misleading, presentations in textbooks and other curriculum materials (Salmon 2000). The Independent Commission on Environmental Education (ICEE) and its successor, the Environmental Literacy Council (ELC), have pursued this concern. ELC's membership is largely discipline-oriented scientists and economists, but also includes an environmental scientist, an environmental educator, a science educator, a secondary school science teacher, and an environmental historian. ICEE/ELC publications to date include content evaluations of widely used K-12 environmental education teaching materials, both supplementary (ICEE 1997) and textbook (ELC 2001). The evaluations were performed by council members, and were based on criteria emphasizing rigorous science, mathematics, and economics. Both evaluations reported a wide range in quality, from very good to seriously flawed. Other ELC products include a teaching resource guide (deBettencourt *et al.* 2000) and a website (www.enviroliteracy.org) that also features teaching resource materials. With National Science Foundation funding, ELC currently is working with the National Science Teachers Association on the development of a high-school environmental science curriculum guide parallel to the American Association for the Advancement of Science's *Project 2061: Science for All Americans* (AAAS 1989). With National Endowment for the Humanities funding, ELC also is in the process of developing humanities-related environmental literacy resources for educators.

Both environmentalists and environmental educators have reacted strongly to accusations of bias and charges of content inadequacy in environmental education, sometimes with ad hominem arguments – i.e., what would you expect from groups and individuals holding vested interests in the overuse, misuse, and abuse of the environment? Environmental educators (Smith 2000; Holsman 2001) have charged that there is at least as much bias in the arguments advanced by the anti-environmental movement as exists in the materials of which they are critical. It remains to be seen whether or not common ground can be reached in the classroom before that happens in society in general. In the meantime, the debate rages on, often with more heat than light.

FINDING A BALANCE?

The environmentalist/anti-environmentalist debate offers a remarkable opportunity for teachers and students to demand clearly stated justifications of these contrasting positions on environmental issues and then to analyze them in class, though neither environmentalists nor anti-environmentalists mention this possibility. For such an analysis to be educationally effective, teachers and students must possess the scientific-technical-economic-political-psychological background, along with the intellect and the will, to understand them. To set the stage properly for learning in this fashion, a number of questions are worthy of investigation by educators. (1) What is the current extent of one-sided endorsement of pro-environmental positions in the classroom? (2) Are there one-sided endorsements of anti-environmental positions in other classrooms, and if so, to what extent? (3) How should educators deal with environmental bias, pro- or anti-, in terms of their own classroom behavior? (4) If there are problems of biased presentation, how might they be resolved? These questions need to be asked and answered *by* professional educators, not *for* professional educators by environmentalists, anti-environmentalists, or anyone else – though their reasoned input should be sought, and could be helpful.

Environmentalism in the classroom remains a bone of contention between and among teachers and other professional educators, environmentalists and anti-environmentalists, political liberals and political conservatives, scientists and economists, and (to some extent) members of the general public. But little effort has been expended to characterize systematically what actually happens in classrooms. One national (US) survey initiated by ELC and NAAEE, and supported by NEETF, drew responses from a random sample of 1505 classroom teachers. It reported that: (1) elementary school teachers teach about the environment more often than do middle school or high school teachers; (2) science teachers teach about the environment more than do teachers of other subjects; (3) about 60 percent of the K-12 teachers in the United States include environmental topics in their curricula; and (4) about half of those who teach about the environment do so, at least in part, to encourage students to protect the environment (Survey Research Center 2000). This is a start, but more research is needed to characterize the field properly. Also needed is research-based information about parental dispositions toward environmental education in its various forms; whether or not parental feelings about teaching about environmental problems and issues in their children's classrooms are parallel to public concerns about environmental quality would be an interesting question to explore.

ENVIRONMENTALLY RESPONSIBLE BEHAVIOR AND ENVIRONMENTAL LITERACY

What, and how, should students be taught so that they actually will behave in an environmentally responsible manner? What, for that matter, do we mean when we talk about "environmentally responsible behavior"? Early on, the frequently accepted answers to both questions were simplistic, and basically authority-oriented: environmentally responsible behaviors are those endorsed by professional resource managers and/or those championed by environmental advocates. It follows that environmental educators should teach students what those behaviors are, and how to achieve them. More recently, it has been argued that more educationally appropriate definitions of environmentally responsible behavior must include ecological knowledge, environmental sensitivity, knowledge of issues, investigation skills, citizenship skills, and feeling of effectiveness (Volk 1993).

"Environmental literacy" is generally seen as the bottom-line goal of environmental education. However, the term is interpreted differently by different people, leading to contrasting expectations of what schools should do to promote it. Educators and non-educators agree that cognitive knowledge is the essential basis of environmental literacy, but some environmental educators identify three additional essential components: skills, as applied to human interactions with the environment; affect, including emotional traits and dispositions toward the environment; and behavior, encompassing activities intended to maintain or improve the quality of the environment (Roth 1992). Defined this way, environmental literacy is close to synonymous with responsible environmental behavior; that is, environmentally literate individuals will, by this definition, behave in an environmentally responsible manner. However, those who are critical of environmental education as currently practiced limit their conception of education for environmental literacy to the acquisition of cognitive knowledge, which they describe as often inadequately presented. They charge that efforts to influence environmental behaviors are typically approached in a biased manner, emotionally and nonsubstantively, and thus have no place in the classroom.

Research in the application of cognitive psychology theories to environmental education has a potential for advancing understanding of the relationships between what we learn, how we learn it, and what difference it makes in terms of our subsequent attitudes and behaviors. Hungerford and Volk (1990) have developed a teaching–learning model that they believe leads to the development of environmentally responsible behaviors. Though this model is verbally endorsed by many environmental educators as an appropriate way to approach the study of the environment in schools, in practice it has not been broadly implemented, probably because it places unusual demands of pedagogic rigor on both teachers and learners. Apparently, the bottom line is that it is much easier to "preach" about positive environmental behaviors than to teach in such a manner that they are developed, internalized, and consistently demonstrated. At the same time, the Hungerford–Volk model has been criticized as being an overt attempt at pro-environmental "brainwashing," though there is little evidence to support that charge.

Little has been done by anyone other than Roth (1992), and Hungerford and his colleagues (1990), to investigate environmental literacy. Little has been done by anyone else to investigate systematically and scientifically approaches to teaching for the development of responsible environmental behaviors. Why is comprehensive environmental literacy so rarely pursued, and teaching for the development of responsible environmental behavior so meagerly implemented? Should teaching for environmental literacy actually be limited to the development of cognitive knowledge? Is teaching for responsible environmental behavior actually equivalent to "brainwashing"? The purposeful application of extensive and intensive straightforward, intelligent, scholarly rigor to such investigations by environmental educators, and professional educators in general, has the potential to advance our understanding and influence subsequent educational practice in these areas.

CONCLUSION: WHAT SCHOOLS SHOULD TEACH

One reason why so many who are not educational professionals have such specific ideas of what schools should teach is that they all went to school sometime or times, someplace or places, and accordingly feel that they possess functional knowledge, perhaps even wisdom, of the purposes of the educational enterprise. People generally believe that schools are places to learn comprehensively, systematically, and efficiently, because schools are the places that most of us first and most intensively became engaged in organized learning and we all know that schooling can, should, and often does work. Moreover, it is obvious to all that in schools young people are gathered in large numbers for the primary purpose of learning; they are in effect a captive audience, and they are developmentally ready to learn at some level whatever it is that professional educators undertake to teach them. All of us agree that the school is an appropriate place for individuals to learn basic skills and specific content. However, we are not in agreement about how we expect schools to deal with values, those things we think are worth taking positions on.

Though we identify the schools we attended as the places where we gained many of the basic skills and much of the cognitive knowledge important to us, it is more difficult to know where we learned what we now think is important enough to value – at home, in the school, from the church, from our friends, colleagues, and casual acquaintances, in the marketplace, from the media, or in some other manner. Probably the answer is by some combination of these, because our values are determined by how we perceive what is going on around us. Should schooling have a specific role in teaching values? From a practical perspective, if those of us who possess either consistent or fragmented value systems – that is, all of us – can influence the educational community (i.e., "educate the educators") there is potential access to large audiences with our value-laden messages. But is it the right thing to do?

It is unrealistic to expect schools to serve solely as value-free dispensers of knowledge for its own sake (Simmons and Volk 2002); the very act of deciding what content, concepts, and processes are important enough to teach is value-laden, and that decision in itself reveals the values of the educator and those who influence the educator. Generally, people understand why schools need to teach mathematics, science, languages, and reading; it is because these give citizens the basic tools necessary to face the realities of modern life. It apparently is not quite so clear that teaching about human interactions with the environment has a similar value.

Generally, people think that everyone should be required to learn why the values they themselves hold are important, regardless of where they learned them or how complete or fragmentary their personal value systems are. There certainly is a need for schools to teach about the values that permeate our society but the problem is what they are, what they mean, where they came from, and how they influence us individually and as a society. That does not mean teaching *for* (i.e., unequivocally endorsing) any particular set of values, but it does mean teaching *about* all of the values, and it may call for more wisdom than most of us possess.

In a pluralistic society there are too many value systems for any of us to master them all, though it is worth the time and effort to become conversant with them and to be able to discuss them rationally with both colleagues and students. It certainly is impossible for educators, who after all are human, to believe in all of them. The following approach, a broadened paraphrase of a short essay (Hug 1977: 73) that first appeared in print more than a quarter of a century ago, is one way that individual educators might approach the problem:

> My suggestion is simply that environmental educators make an effort [in their teaching behavior] to clarify the distinct roles [of the proponents of all environmental perspectives]. At every opportunity, we should emphasize the neutral nature of environmental education activity. Strong advocates are all around us, each using the techniques

> of persuasion and propaganda to build their constituencies. We must, ourselves, be familiar with all sides, stand firm for each advocate's right to be heard, and provide a rational stage for informed debate.

To the extent that individual educators are able to achieve such idealized teaching behavior, the problem will be resolved. But if they are unable to do it, the credibility of education as a profession is compromised. In such cases, teachers and other educators will uncritically rely on the readily available inputs of those whose interests in education are secondary to their political, economic, social, or environmental agendas.

ACKNOWLEDGMENTS

The author gratefully acknowledges comments and suggestions offered on various drafts of this chapter by Harold R. Hungerford, Charles E. Roth, unnamed reviewers, and the editors of this volume.

REFERENCES

American Association for the Advancement of Science (AAAS). (1989). *Project 2061: Science for All Americans*. New York, NY: Oxford University Press.

Carson, R. (1962). *Silent Spring*. Boston, MA: Houghton-Mifflin.

Coyle, K. (1996). *Briefing Report on the Emerging Conservative Opposition to Environmental Education: Some Implications for Environmental Grant-making*. Washington, DC: National Environmental Education and Training Foundation.

deBettencourt, K. B., Feeney, M., Barone, A. N., and White, K. (2000). *Environmental Connections: A Teacher's Guide to Environmental Studies*. Dubuque, IA: Kendall-Hunt Publishing Company.

Disinger, J. F. (1978). *Alliance Affiliate Activities: Non-Governmental Organizations in Environmental Education*. Columbus, OH: ERIC/SMEAC.

(2001). Tensions in environmental education: yesterday, today, and tomorrow. In *Essential Readings in Environmental Education*, 2nd edn, ed. H. R. Hungerford, W. J. Bluhm, T. R. Volk, and J. M. Ramsey. Champaign, IL: Stipes Publishing L.L.C., pp. 1–12.

Dunlap, R. E. (1987a). Polls, pollution and politics revisited: public opinion on the environment in the Reagan era. *Environment*, **29**(6), 6–11, 32–37.

(1987b). Response to McCloskey and Udall. *Environment*, **29**(9), 2.

Environmental Literacy Council (ELC). (2001). *Science for Environmental Literacy: A Review of Advanced Placement Environmental Science Textbooks*. Washington, DC: ELC.

Federal Interagency Committee on Education, Subcommittee on Environmental Education (FICE/SEE). (1976). *Fundamentals of Environmental Education*. Washington, DC: US Department of Health, Education and Welfare, Office of the Assistant Secretary for Education.

Funderburk, R. S. (1948). *The History of Conservation Education in the United States*. Nashville, TN: George Peabody College for Teachers.

Hays, S. P. (1985). *Beauty, Health, and Permanence: Environmental Politics in the United States, 1955–1985*. New York, NY Cambridge University Press.

Holsman, R. H. (2001). Viewpoint: the politics of environmental education. *Journal of Environmental Education*, **32**(2), 4–7.

Hug, J. (1977). Two hats. In *Report of the North American Regional Seminar on Environmental Education*, ed. J. Aldrich, A. Blackburn, and G. Abel, Columbus, OH: SMEAC/IRC, p. 73.

Hungerford, H. R. and Volk, T. L. (1990). Changing learner behavior through environmental education. *Journal of Environmental Education*, **21**(3), 8–21.

Independent Commission on Environmental Education. (1997). *Are We Building Environmental Literacy?* Washington, DC: ICEE.

Jeske, W. E. (1978). Subcommittee on Environmental Education, Federal Interagency Committee on Education. In *Environmental Education Activities of Federal Agencies*, ed. J. F. Disinger. Columbus, OH: ERIC/SMEAC, pp. 159–163.

Lucas, A. M. (1972). Environment and environmental education: conceptual issues and curriculum implications. Ph.D. dissertation, The Ohio State University, Columbus. *Dissertation Abstracts International*, **33**, 6064-A.

McSwain, J. (1978). Alliance for Environmental Education. In *Alliance Affiliate Activities: Non-Governmental Organizations in Environmental Education*, ed. J. F. Disinger. Columbus, OH: ERIC/SMEAC, pp. 8–13.

Myshak, R. J. (1978). Conservation Education Association. In *Alliance Affiliate Activities: Non-Governmental Organizations in Environmental Education*, ed. J. F. Disinger. Columbus, OH: ERIC/SMEAC, pp. 60–61.

NAAEE. (1996). *Environmental Education Materials: Guidelines for Excellence*. Rock Spring, GA: North American Association for Environmental Education.

(1997, 1998). *The Environmental Education Collection: A Review of Resources for Educators* (3 volumes). Rock Spring, GA: North American Association for Environmental Education.

Roth, C. E. (1978). Off the merry-go-round and on to the escalator. In *From Ought to Action in Environmental Education*, ed. W. B. Stapp. Columbus, OH: SMEAC/IRC, pp. 12–22.

(1992). *Environmental Literacy: Its Roots and Directions in the 1990s*. Columbus, OH: ERIC/CSMEE.

Russell, H. R. (1978). American Nature Study Society. In *Alliance Affiliate Activities: Non-Governmental Organizations in Environmental Education*, ed. J. F. Disinger. Columbus, OH: ERIC/SMEAC, pp. 45–46.

Salmon, J. (2000). Are we building environmental literacy? *Journal of Environmental Education*, **31**(4), 4–10.

Sanera, M. and Shaw, J. S. (1996). *Facts, Not Fear: A Parent's Guide to Teaching Children About the Environment*. Washington, DC: Regnery Publishing.

Simmons, B. and Volk, T. (2002). Environmental educators: a conversation with Harold Hungerford. *Journal of Environmental Education*, **34**(2), 5–8.

Smith, G. A. (2000). *Defusing environmental education: an evaluation of the critique of the environmental education movement*. Center for Education Research, Analysis, and Innovation, University of Wisconsin-Milwaukee.

Stapp, W. B., Bennett, D., Bryan, W., Fulton, J., MacGregor, J., Nowak, P., Swan, J., Wall, R., and Havlick, S. (1969). The concept of environmental education. *Journal of Environmental Education*, **1**(1), 30–31.

Survey Research Center. (2000). *Environmental Studies in the Classroom: A Teacher's View*. College Park: University of Maryland.

Swan, M. (1975). Forerunners of environmental education. In *What Makes Education Environmental?*, ed. N. McInnis and D. Albrecht. Louisville, KY: Data Courier, Inc., pp. 4–20.

UNESCO. (1972). *Environmental Education and Training: Suggestions Developed by the Secretary General of the United Nations Conference on the Human Environment* (Stockholm, Sweden, 1972). Paris, France: UNESCO.

(1978). *Final Report, Intergovernmental Conference on Environmental Education* (Tbilisi, Georgia, USSR, 1977). Paris: UNESCO.

US Public Law 91-190. (1970). The National Environmental Policy Act of 1969.

US Public Law 91-516. (1970). The Environmental Quality Education Act of 1970.

US Public Law 101-619. (1990). The National Environmental Education Act of 1990.

Volk, T. (1993). Integration and curriculum design. In *Environmental Education Teacher Resource Handbook*, ed. R. J. Wilke. Thousand Oaks, CA: Corwin Press, pp. 45–75.

Western Regional Environmental Education Council (WREEC). (1975). *Project Learning Tree*. Washington, DC: American Forest Institute.

Part III Assessing changing perspectives of ecology and education

This section provides educators with some possible guidelines and tools to assess the education and science components of environmental education programs, resources, and issues. Deborah Simmons introduces the reader to the purpose and process involved in developing a national framework for environmental education in the United States. Simmons provides an overview of the project, along with issues related to the development and use of the frameworks. David Haury extends this work into practice by providing an approach to assessing the educational value of resources provided by non-formal groups or agencies outside of school. Haury uses the "practical framework" of curriculum inquiry to develop a framework for assessing resources in the context of the instructional approach, aims, and content of subject matter, learners, and priorities or values of school and community. The chapter also highlights other assessment resources available for educators. The section concludes with a framework to investigate and evaluate the science involved in environmental issues developed through research and practice by Stein Dankert Kolstø. Kolstø provides an overview on how students assess the complexity of environmental issues and the science involved. The framework for developing knowledge about science is compared with other approaches to developing thoughtful decision-making by students.

DEBORAH A. SIMMONS

8

Developing guidelines for environmental education in the United States: the National Project for Excellence in Environmental Education

INTRODUCTION

Environmental education in the United States has always been a grassroots endeavor, characterized by thousands of educators working in schools, colleges, nature centers, zoos, museums, government agencies, and non-governmental organizations. It must also be said that formal education in the United States is decentralized. There is no national curriculum; there are no national exams. Each state determines how schools will function. Some states have a state-mandated curriculum and statewide adoption of textbooks. Others allow each school district to determine its own curriculum and select its own teaching materials. Consequently, when education is discussed, let alone environmental education in the United States, that discussion must be framed in terms of generalities and commonly accepted notions.

Although at times it may seem difficult for everyone to agree upon the exact wording of a definition, the practice of environmental education in the United States is characterized by some essential elements (Disinger and Monroe 1994):

- Environmental education is based in knowledge about ecological and social systems. It draws on and integrates knowledge from disciplines that span the natural sciences, social sciences, and humanities.
- Environmental education considers humans and their creations to be a part of the environment. Along with biological and physical phenomena, environmental education considers social, economic, political, technological, cultural, historical, moral, and aesthetic aspects of environmental issues.

- Environmental education emphasizes the critical thinking and problem-solving skills needed for informed personal decisions and public action.
- Environmental education emphasizes the role of attitudes, values, and commitments in shaping environmental issues. It acknowledges that environmental issues are not strictly scientific in nature. Recognizing the feelings, values, attitudes, and perceptions at the heart of environmental issues is an essential step in understanding them, and a precursor to accepting responsibility for exploring, analyzing, and resolving them.

Even with descriptions of the elements of environmental education as proposed by Disinger and Monroe, many in the field felt that environmental education needed generally agreed upon frameworks for practice. This chapter describes the efforts of the National Project for Excellence in Environmental Education to develop frameworks and concludes with a discussion of issues related to the process.

NATIONAL PROJECT FOR EXCELLENCE IN ENVIRONMENTAL EDUCATION

Environmental education in the United States has continuously grappled with the question of what it means to be an environmentally literate citizen and how to design and implement effective, comprehensive environmental education programs. In 1993 the North American Association for Environmental Education (NAAEE) initiated the National Project for Excellence in Environmental Education in an attempt to address these critical issues. To some extent the project began as a response to the education reform movement of the late 1980s and early 1990s within the United States.

The calls for national education standards were first heralded with the publication of *A Nation at Risk* (National Commission on Excellence in Education 1983). It became common to call into question the very structure of American education. In at least a partial response to the concerns raised in *A Nation at Risk*, each of the core curriculum areas (i.e., science, geography, mathematics, English-language arts, history, civics, etc.) developed a set of voluntary national standards. These standards, many of which have been adapted or adopted at the state level, delineated the knowledge and skill bases of their respective fields. They are designed to define what students should know and be able to do in order to be considered geographically literate, scientifically literate, or mathematically literate by the time they graduate from secondary school.

The National Project was designed to establish guidelines for the development of balanced, accurate, and comprehensive environmental education programs and to identify and provide examples of high quality environmental education practice. The Project initiated five interrelated efforts: (1) *Environmental Education Materials: Guidelines for Excellence* (NAAEE 1996); (2) *The Environmental Education Collection – A Review of Resources for Educators* (Simmons and Vymetal-Taylor 1997, 1998a, 1998b); (3) *Excellence in Environmental Education – Guidelines for Learning (K-12)* (NAAEE 1999); (4) *Guidelines for the Initial Preparation of Environmental Educators* (NAAEE 2000); and (5) *Non-formal Environmental Education Programs: Guidelines for Excellence* (2004).

The National Project is sponsored by NAAEE and administered by Northern Illinois University. Primary funding came from the US Environmental Protection Agency. Because of the funding source and specific issues related to education reform efforts in the United States, the various Guidelines documents were developed primarily for a US audience.

Environmental education materials: guidelines for excellence

The *Environmental Education Materials: Guidelines for Excellence* aims to help developers of activity guides, lesson plans, and other instructional materials to produce high-quality products, and to provide educators with a tool to evaluate the wide array of available environmental education materials. Starting in the mid-1990s, environmental education was criticized for being biased, most notably by Michael Sanera and the George C. Marshall Institute. It should be noted, the development of *Environmental Education Materials: Guidelines for Excellence*, was not, as some believe, in response to this criticism. The scope of the National Project for Excellence in Environmental Education was established in early 1994 (Simmons 1995). The Sanera and Shaw volume, *Facts, Not Fear: Teaching Children about the Environment*, was not published until 1996.

Taken as a whole, the *Materials Guidelines* offers a way of judging the relative merit of different materials, a set of benchmarks to aim for in developing new materials, and a set of ideas about what a well-rounded environmental education curriculum might be like. It is not reasonable, however, to expect that all environmental education materials will follow all of the guidelines. If used as a touchstone, the *Materials*

Guidelines can point out limitations that instructors can compensate for in the way they use the instructional materials. For example, a set of materials might not present differing viewpoints, as outlined in one of the guidelines. This shortcoming does not necessarily mean that the materials should not be used. An educator could work the particular materials into a larger set of activities that explored different viewpoints and helped learners discern opinion and bias in individual presentations of the issue.

Development of Materials Guidelines

In an effort to ensure that the *Materials Guidelines* reflected a widely shared understanding of environmental education, they were developed though a nation-wide process of review and comment. A ten-person writing team comprised of environmental education professionals from a variety of backgrounds and organizational affiliations was formed. The writing team took on the challenge of turning ideas about quality, gleaned from the environmental education literature and instructional materials development research literature, into a detailed outline. This outline, along with successive drafts of the guidelines, was circulated widely.

Short articles announcing the availability of each draft and inviting participation in the review process were published in newsletters around the United States (e.g., National Science Teachers Association, National Council for Social Studies, NAAEE, Ecological Society of America, Wildlife Society, state environmental education organizations, environmental organizations). Presentations were made at state, regional, and national environmental education, formal education, and environmental meetings to publicize the effort and to encourage participation. Any individual or organization wishing to participate in the process was encouraged to do so (that is, participation was not restricted in any way). Efforts to publicize the project and the availability of review drafts were made continuously throughout the process. As individuals and organizations volunteered to review the draft documents, a mailing list was created. Each time a new draft was developed, it was sent to everyone on the mailing list. Although comments were received from individuals from 30 countries, the vast majority of comments were from the United States.

As comments were received, they were entered verbatim into a master database. General comments were listed together, and comments relating to specific sections of the draft (e.g., introduction, key characteristics, glossary) were grouped together. This allowed the writing team to consider each comment individually and within the context offered by the draft document. All comments were included in the compilation, from those that suggested word changes to major additions to the complete reorganization of the document. Changes were made in the successive drafts based on an analysis of these comments. Where conflicting views could not be reconciled, revisions were made, in most instances, to reflect the preponderance of opinions expressed. By the end of the review process, over 1000 practitioners and scholars in the field (e.g., classroom teachers, education administrators, environmental scientists, curriculum developers, non-formal environmental educators) participated in the review and development of this document. Review comments were used not only to test and revise the initial outline, but also to develop every detail of the document from overall structure to examples, and glossary terms to references.

Structure and application of Materials Guidelines

The guidelines are organized around what reviewers agreed were six key characteristics of high-quality environmental education material: fairness and accuracy, depth, emphasis on skill building, action orientation, instructional soundness, and usability. For each of these characteristics, specific guidelines are listed that support the implementation of the key characteristics (Table 8.1). Finally, each of the guidelines is accompanied by several indicators listed under the heading "What to Look For." These indicators suggest ways of gauging whether the materials being evaluated or developed reflect the notions expressed by the guidelines. These indicators are simply clusters of attributes that might help determine whether the guideline is embodied in the materials being reviewed or developed (Table 8.2).

Educators need good-quality environmental education materials, but with literally thousands of products to select from, deciding which materials best meet their needs can be overwhelming. The National Project used the *Materials Guidelines* as the criteria for review to develop *The Environmental Education Collection: A Review of Resources for Educators.* A broad range of educational materials (e.g., curriculum guides and lesson packets, CD-ROMs, computer software, laser discs, video tapes, etc.) was included in the review. Each set of instructional materials was examined by at least seven professionals. Over the course of the project, panels of educators and content specialists examined literally hundreds of educational materials. NAAEE published their reviews in

Table 8.1. *Summary of* Environmental Education Materials: Guidelines for Excellence (NAAEE 1996 (2004a))

(1) **Fairness and accuracy:** EE materials should be fair and accurate in describing environmental problems, issues, and conditions, and in reflecting the diversity of perspectives on them.

- 1.1 Factual accuracy
- 1.2 Balanced presentation of differing viewpoints and theories
- 1.3 Openness to inquiry
- 1.4 Reflection of diversity

(2) **Depth:** EE materials should foster awareness of the natural and built environment, an understanding of environmental concepts, conditions, and issues, and an awareness of the feelings, values, attitudes, and perceptions at the heart of environmental issues, as appropriate for different developmental levels.

- 2.1 Awareness
- 2.2 Focus on concepts
- 2.3 Concepts in context
- 2.4 Attention to different scales

(3) **Emphasis on skills building:** EE materials should build lifelong skills that enable learners to address environmental issues.

- 3.1 Critical and creative thinking
- 3.2 Applying skills to issues
- 3.3 Action skills

(4) **Action Orientation:** EE materials should promote civic responsibility; encourage learners to use their knowledge, personal skills, and assessments of environmental issues as a basis for environmental problem solving and action.

- 4.1 Sense of personal stake and responsibility
- 4.2 Self-efficacy

(5) **Instructional soundness:** EE materials should rely on instructional techniques that create an effective learning environment.

- 5.1 Learner-centered instruction
- 5.2 Different ways of learning
- 5.3 Connection to learners' everyday lives
- 5.4 Expanded learning environment
- 5.5 Interdisciplinary

5.6 Goals and objectives
5.7 Appropriateness for specific learning settings
5.8 Assessment

(6) **Usability:** EE materials should be well designed and easy to use.

6.1 Clarity and logic
6.2 Easy to use
6.3 Long-lived
6.4 Adaptable
6.5 Accompanied by instruction and support
6.6 Make substantiated claims
6.7 Fit with national, state, or local requirements

three resource guides. A fourth guide, developed using the same procedures, was published by World Wildlife Fund (Braus 1998).

In addition to being used as a review instrument, the guidelines can be used in the development and revision of various instructional materials (e.g., CD-ROMs, web-based instructional units, public television study guides, nationally disseminated curriculum guides, lesson packets designed for museums or zoos, and teacher-created units).

Issues related to Materials Guidelines

It became clear during the development process that the *Materials Guidelines* needed to address two issues specific to environmental education: (1) education versus advocacy and (2) environmental action. Of particular interest is the focus on factual accuracy, balance, and openness to inquiry. Environmental education often deals with topics and issues that are contentious. Consequently, it was felt that the *Materials Guidelines* should address different viewpoints and explanations, encourage learners to explore differing perspectives and form their own opinions, and acknowledge that feelings, experiences, and attitudes shape perceptions. In addition, the guidelines suggest that environmental education should challenge learners to think creatively and critically, give learners the opportunity to arrive at their own conclusions, help learners gain basic skills needed to participate in resolving environmental issues, encourage learners to examine the possible consequences of their behaviors and evaluate choices, and aim to strengthen learners' perceptions of their ability to influence the outcome of a situation. In focusing on these characteristics, the *Materials*

Table 8.2. *Key Characteristic No. 4 from* Environmental Education Materials: Guidelines for Excellence (NAAEE 1996 (2004a))

Key characteristic No. 4 Action Orientation

Environmental education materials should promote civic responsibility, encouraging learners to use their knowledge, personal skills, and assessments of environmental issues as a basis for environmental problem solving and action.

(4.1) **Sense of personal stake and responsibility.** Materials should encourage learners to examine the possible consequences of their behaviors on the environment and evaluate choices they can make which may help resolve environmental issues.

What to look for

- Materials promote intergenerational and global responsibility, linking historical and current actions with future and distant consequences.
- Learners are provided with opportunities to reflect on the effects of their actions and to sort out their opinions about what, if anything, they should do differently.
- Materials contain examples of people of different ages, races, genders, cultures, and education and income levels who have made a difference by taking responsible action.
- Materials convey the idea that many individual actions have cumulative effects, both in creating and addressing environmental issues.

(4.2) **Self-efficacy.** Materials should aim to strengthen learners' perception of their ability to influence the outcome of a situation.

What to look for:

- Materials challenge learners to apply their thinking and act on their conclusions.
- Materials include a variety of individual and community strategies for citizen involvement and provide learners with opportunities to practice these strategies through projects they generate individually in their school or in the larger community.
- There are examples of successful individual and collective actions. Learners are encouraged to examine what made these actions successful. (Where actions were not successful, students are encouraged to examine the reasons for failure.)
- Learners are encouraged to share the results of their actions with peers and other interested people.

Guidelines specifically and consciously encourage the development of citizenship skills and promote civic responsibility *without endorsing any particular course of action.*

Although the importance of citizenship skills was endorsed by those who commented when the guidelines were being developed, it became clear when reviewing published materials that the instructional components related to the key characteristics of "Emphasis on Skills Building" and "Action Orientation" were often lacking. Many environmental education materials emphasize awareness, appreciation, and knowledge, without an accompanying focus on developing skills and commitment to action.

Some environmental educators have suggested that developing citizenship skills may be unrealistic given the format of many instructional materials. That is, environmental education materials are often designed for use within relatively short lessons or units (one or two hours) and not as part of a larger scope and sequence. It has been argued that, since citizenship skills cannot be developed within this time frame, key characteristics and guidelines related to citizenship should be dropped from the overall framework. In other words, some are suggesting that the framework should be modified to mirror the actual practice reflected in these materials. The National Project has taken the stance that developing citizenship skills is an important component of environmental education (as reflected in the comments received during the development of the *Materials Guidelines*). It is hoped that by including these key characteristics more materials that support a comprehensive vision of environmental education will be developed.

Excellence in Environmental Education: Guidelines for Learning (K-12)

The *Excellence in Environmental Education: Guidelines for Learning (K-12)*, published in 1999, provides links between the standards-based core curriculum and environmental education. The *Guidelines for Learning* was developed to provide students, parents, educators, administrators, policymakers and the public a set of common voluntary guidelines for environmental education. The guidelines support state and local environmental education efforts by defining the aims of environmental education as well as:

- setting expectations for performance and achievement in fourth, eighth, and twelfth grades;

- suggesting a framework for effective and comprehensive environmental education programs and curricula; and
- demonstrating how environmental education can be used to meet standards set by the traditional disciplines and giving students opportunities to synthesize knowledge and experiences across disciplines.

Although obviously written to address the needs of specific discipline-based areas, the standards for core curriculum do, to one degree or another, address environmental education interests. Taken singly, the standards of any one discipline allow for environmental learnings. For example, ecological knowledge such as the components of the earth's physical systems (the atmosphere, lithosphere, hydrosphere, and biosphere), how earth–sun relations affect conditions on earth and the physical characteristics of places (e.g., landforms, bodies of water, soil, vegetation, and weather and climate) are included within *Geography for Life: National Geography Standards* (National Geographic Research and Exploration 1994).

Conversely, environmental education programs can be used to meet the discipline-based standards. Because environmental education is by its very nature interdisciplinary, it can help students meet the high standards set by traditional school disciplines (e.g., science, civics, geography, history). Although environmental education can effectively and efficiently facilitate the learning of specific concepts and process skills, it also provides an often missed opportunity for the synthesis of materials that crosses disciplinary boundaries, connecting "learning" to create a whole. The explicit focus of environmental education on the integration of knowledge and skills is one of the primary distinguishing factors between it and a traditional view of curricular disciplines. Environmental education has the potential of linking the K-12 curriculum by providing the opportunity to meet the requirements of the core disciplines and creating a comprehensive and cohesive program of study.

Environmental education, however, also has an essential purpose in and of itself: *environmental literacy*. The *Guidelines for Learning* provides a vision of environmental literacy that acknowledges that a knowledgeable, skilled, and active citizenry is key to preventing and resolving current and future environmental problems. The guidelines offer a framework for environmental education that examines the relationship between the environment and quality of life, and uses an interdisciplinary approach emphasizing thinking and action skills central to environmental literacy. Environmental literacy must be a goal of our society; consequently, environmental education must play an integral role throughout our educational system - at the national level, state level, and in each and every classroom. The guidelines is aimed at providing a series of tools that might help educators develop effective, locally relevant environmental education programs leading to environmental literacy.

Development of Guidelines for Learning

A conscious effort was made to model the *Guidelines for Learning* after the national standards published by the various discipline-based groups (e.g., National Council of Geographic Education, Center for Civic Education, National Council for the Social Studies). Modeling the *Guidelines for Learning* after the various national standards aided the process in two specific ways: (1) it ensured that the structure and format would be familiar to many educators and (2) an additional check regarding the age appropriateness of concepts and skills was achieved. The guidelines were designed to provide teachers and other educators with the tools to support the development of coherent, comprehensive environmental education programs that meet high education standards.

The writing process followed the general protocols as used in the development of the *Materials Guidelines*. By the time they were published, over 2500 individuals and organizations had commented. Analysis of comments also followed a similar procedure as used with the *Materials Guidelines*.

Structure and application of Guidelines for Learning

The guidelines are organized into four strands, each of which represents a broad aspect of environmental education: questioning and analysis skills, knowledge of environmental processes and systems, skills for understanding and addressing environmental issues, and personal and civic responsibility (Table 8.3). Individual guidelines and performance measures are suggested for each of three grade levels - fourth, eighth, and twelfth. Each guideline is also cross-referenced to related discipline-based standards in fields including English-language arts, mathematics, science, and the social studies.

When the development process began, there was a hope that the guidelines might influence the development of various state standards. Unfortunately, by the time the *Guidelines for Learning* were published in 1999, most of the states had already produced their student academic

Table 8.3. *Strands in* Excellence in Environmental Education: Guidelines for Learning (K-12) (NAAEE 1999 (2004b))

Strand 1: Questioning and analysis skills

Environmental literacy depends on learners' ability to ask questions, speculate, and hypothesize about the world around them, seek information, and develop answers to their questions. Learners must be familiar with inquiry, master fundamental skills for gathering and organizing information, and interpret and synthesize information to develop and communicate explanations.

Strand 2: Knowledge of environmental processes and systems

An important component of environmental literacy is understanding the processes and systems that comprise the environment, including human systems and influences. That understanding is based on knowledge synthesized from across traditional disciplines. The guidelines in this section are grouped in four sub-categories:

2.1 The Earth as a physical system
2.2 The living environment
2.3 Humans and their societies and
2.4 Environment and society

Strand 3: Skills for understanding and addressing environmental issues

Skills and knowledge are refined and applied in the context of environmental issues. These environmental issues are real-life dramas where differing viewpoints about environmental problems and their potential solutions are played out. Environmental literacy includes the abilities to define, learn about, evaluate, and act on environmental issues. In this section, the guidelines are grouped in two sub-categories:

3.1 Skills for analyzing and investigating environmental issues; and
3.2 Decision-making and citizenship skills.

Strand 4: Personal and civic responsibility

Environmentally literate citizens are willing and able to act on their own conclusions about what should be done to ensure environmental quality. As learners develop and apply concept-based learning and skills for inquiry, analysis, and action, they also understand that what they do individually and in groups can make a difference.

standards. However, in some cases (Hawaii and Wisconsin in particular), state educators were aware of the development of the guidelines throughout the comment and review process. Consequently, these state academic standards do reflect the understandings and skills needed for environmental literacy. In other cases, state educators recognized that, although it was too late to influence the writing of their state academic standards, the guidelines offered an important tool for showing the relationship between environmental education and the K-12 curriculum. Educators in Kentucky and Illinois have correlated the guidelines to their state academic standards and published the results on the Internet. Several other states are in the process of conducting a similar analysis.

Although the *Guidelines for Learning* were developed, in part, to address education reform issues, the larger purpose was to provide a framework for a comprehensive environmental education curriculum. To this end, curriculum developers have begun to use the guidelines in their efforts. Additionally, some organizations have begun to highlight the relationship between their materials and the guidelines. As an example, each lesson in the *Project Learning Tree* (American Forest Foundation 1997) activity guide has been correlated to the *Guidelines for Learning.*

Issues related to the Guidelines for Learning

Academic standards aim to describe what a student should know and be able to do at a particular point in their schooling. Academic standards focus on cognition and skill development. Environmental education has traditionally included the affective dimension and the need for the development of environmental awareness as goals. Consequently, a major issue in the development of the *Guidelines for Learning* centered on how or if the affective dimension should be addressed. Although it was recognized that affect plays an important role in learning, writing guidelines specific to affect did not seem appropriate. However, guidelines are included related to developing the knowledge base and the skills needed to analyze attitudes, values, and beliefs.

A related issue centered on defining what was central to environmental education and environmental literacy. In writing guidelines a decision had to be made regarding which concepts and skills are critical to environmental literacy *per se* and which are a prerequisite for the concepts and skills that, in the end, are critical to environmental literacy. Similarly, some argued that the scope of the guidelines was too narrow. Some believed that the guidelines should address environmental literacy components appropriate for non-formal education settings and adult learners. Others felt that the guidelines should focus on the use of out-of-classroom environments, in particular natural environments, in environmental education. Again, although these issues are important to environmental education, the *Guidelines for Learning* were designed to correspond to the K-12 standards-based curriculum.

Guidelines for the Initial Preparation of Environmental Educators

The *Guidelines for the Initial Preparation of Environmental Educators* (2004c) offers a set of recommendations about the basic understandings and abilities educators need in order to provide high-quality environmental education. The guidelines were designed to apply:

- within the context of preservice teacher education programs and environmental education courses offered to students with varied backgrounds such as environmental studies, geography, liberal studies, or natural resources;
- to the preparation of instructors who will work in both formal and non-formal educational settings, offering programs at the prekindergarten through 12th grade levels;
- to those preparing to be full-time environmental educators and those for whom environmental education will be among other responsibilities or integrated within the curriculum.

Development of Initial Preparation Guidelines

As with the two preceding guidelines documents, *Guidelines for the Initial Preparation of Environmental Educators* was developed through a process of review and comment. Substantial research related to the preparation of environmental educators was conducted. In addition, the education literature was reviewed with particular attention paid to preservice teacher preparation. This research was used by the writing team as a springboard for developing the first full outline of the guidelines. That outline, along with subsequent drafts, was sent out for national review by the National Project staff. Over the span of two years, over 750 individuals and organizations commented on the draft guidelines. Comments were received from environmental educators from a wide variety of settings, including non-formal institutions, colleges and universities, government agencies, and K-12 schools. Given the nature of this set of guidelines, faculty with expertise in environmental studies, natural resources, and teacher preparation from a wide variety of colleges and universities were particularly active in the review process.

Structure and application of the Initial Preparation Guidelines

The guidelines are organized around six themes: environmental literacy, foundations of environmental education, professional responsibilities of the environmental educator, planning and implementing environmental education programs, fostering learning, and assessment and evaluation (Table 8.4). Each theme describes a knowledge or skill area that should be included in the pre-service preparation of an environmental educator. Under each theme, general guidelines further articulate the knowledge and skills that must be mastered to gain competency in that area. Several indicators that suggest ways of assessing the abilities of new environmental educators accompany each guideline.

The guidelines for teacher education and professional development in environmental education are integrally related to the *Guidelines for Learning*. Describing what students should know and be able to do as environmentally literate citizens determines, to some degree, what educators need to know and be able to do. But being an effective environmental educator requires more than competency with a specific set of knowledge and skills. Educators design programs and curricula, teach students with diverse backgrounds, maintain educational policies, and assess student learning.

Guidelines for the Initial Preparation of Environmental Educators was designed to help those who prepare teachers and other environmental educators develop courses that address environmental education concerns. This is particularly important given that McKeown-Ice (2000: 11) found that "Pre-service teacher education programs are not systematically preparing future teachers to effectively teach about the environment." The guidelines can be used in undergraduate and graduate education courses and as a template for designing the course itself. For example, the guidelines were used as the framework for the development of an on-line course, *Fundamentals of Environmental Education*, produced by the University of Wisconsin at Stevens Point (see http://www.eetap.org/html/online_ee_course.php). The guidelines can also be used as a self-assessment tool for students and instructors.

Over the last several years, a number of states have begun to draft environmental educator certification programs. Modeled after programs that certify individuals in professions such as forestry, wetland ecology, and planning, these state initiatives aim to both improve the

Table 8.4. *Summary of* Guidelines for the Initial Preparation of Environmental Educators (NAAEE 2004c)

Theme No. 1 Environmental literacy

Educators must be competent in the skills and understandings outlined in *Excellence in Environmental Education: Guidelines for Learning (K-12).*

1.1	Questioning and analysis skills
1.2	Knowledge of environmental processes and systems
1.3	Skills for understanding and addressing environmental issues
1.4	Personal and civic responsibility

Theme No. 2 Foundations of environmental education

Educators must have a basic understanding of the goals, theory, practice, and history of the field of environmental education.

2.1	Fundamental characteristics and goals of environmental education
2.2	How environmental education is implemented
2.3	The evolution of the field

Theme No. 3 Professional responsibilities of the environmental educator

Educators must understand and accept the responsibilities associated with practicing environmental education.

3.1	Exemplary environmental education practice
3.2	Emphasis on education, not advocacy
3.3	Ongoing learning and professional development

Theme No. 4 Planning and implementing environmental education programs

Educators must combine the fundamentals of high-quality education with the unique features of environmental education to design and implement effective instruction.

4.1	Knowledge of learners
4.2	Knowledge of instructional methodologies
4.3	Planning for instruction
4.4	Knowledge of environmental education materials and resources
4.5	Technologies that assist learning
4.6	Settings for instruction
4.7	Curriculum planning

Theme No. 5 Fostering learning

Educators must enable learners to engage in open inquiry and investigations, especially when considering environmental issues that are controversial and require students to seriously reflect on their own and others' perspectives.

5.1 A climate for learning about and exploring the environment
5.2 An inclusive and collaborative learning environment
5.3 Flexible and responsive instruction

Theme No. 6 Assessment and evaluation

Environmental educators must posses the knowledge and commitment to make assessment and evaluation integral to instruction and programs.

6.1 Learner outcomes
6.2 Assessment that is part of instruction
6.3 Improving instruction

---

practice of environmental education and elevate the profession. Four states (Utah, Kentucky, Texas, and Georgia) are in the process of designing and implementing environmental education certification programs that assess educator competency as a means of determining certification. Each of these states used the *Initial Preparation Guidelines* as the framework for the environmental educator competencies. The North American Association for Environmental Education is currently examining the possibility of developing a national environmental educator certification program based on competencies.

Public K-12 teachers are certified by the states. To be certified, these teachers must have met specific sets of standards. Similarly, many colleges of education in the United States are accredited by the National Council for Accreditation of Teacher Education (NCATE). NCATE has established sets of standards that are designed to improve the quality of teacher preparation. Currently, NAAEE is working with NCATE to consider how environmental education might be incorporated into these standards. The first step will be to revise the *Guidelines for the Initial Preparation of Environmental Educators* to fit within the NCATE structure. This process is expected to take approximately two years.

Issues related to the Initial Preparation Guidelines

The introduction to the *Guidelines for the Initial Preparation of Environmental Educators* addresses the need to prepare both formal and non-formal environmental educators. Perhaps the biggest criticism of the *Initial Preparation Guidelines*, identified during the development process and implementation workshops, has centered on those themes that focus on education practice. Some environmental educators argued that formal classroom teachers do not need to develop additional competencies related to instruction (e.g., "Fostering Learning"). They suggested that to be most meaningful to formal educators, the guidelines should focus only on those aspects of education unique to environmental education (i.e., "Environmental Literacy," "Foundations of Environmental Education"). Similarly, others argued that for non-formal environmental educators the guidelines focused far too much on education (e.g., "Planning and Implementing Environmental Education Programs") competencies. There was a sense that, particularly for non-formal educators, too much terminology specific to education was used. These individuals argued that, by including guidelines related to knowledge of learners, instructional methodologies, and planning for instruction, the *Initial Preparation Guidelines* became inaccessible to non-formal environmental educators who do not have an education background. There are plans to revise all of the guidelines documents in 2004. Attention will be paid to this issue at that time.

Non-formal environmental education programs: guidelines for excellence

Non-formal environmental education programs are extremely diverse in their settings and in their target audiences. Community-based groups, service organizations, boys' and girls' clubs, Elderhostels, parks, nature centers, zoos, museums, 4-H (an agricultural based youth organization), resident centers, scouting organizations, ecotourism operators, etc., all may be involved in non-formal environmental education. The overall goal of the *Non-formal Program Guidelines* is to provide these groups with tools to develop high quality environmental education programs and to improve existing ones. To assure that the newest entry in the *Guidelines for Excellence* series reflects a widely shared understanding of environmental education, the writing team consisted of professionals from a variety of backgrounds and organizational affiliations.

The current draft of *Non-formal Environmental Education Programs: Guidelines for Excellence* (2004d) identifies six key characteristics, or components, of high-quality programs: needs assessment, organizational needs and capacities, program scope and structure, program delivery resources, program quality and appropriateness, and evaluation (Table 8.5). For each of the characteristics, guidelines are then listed for non-formal program designers to consider. Finally, indicators are provided for each guideline to assess if the non-formal programs have any of the key characteristics. Indicators are simply clusters of attributes that, taken together, describe the individual guidelines.

The guidelines also specifically advocate the development of partnerships as a means of improving the quality and efficiency of environmental education programs. Collaboration between formal and non-formal sectors is becoming more common and widespread, from local nature centers working with school systems to federal agencies working with businesses and universities. This collaboration is essential to increase the quality, reach, and impact of environmental education programs.

It is hoped that non-formal environmental educators will utilize this set of guidelines as an overall framework for program development while creating comprehensive and cohesive environmental education programs that also address issues related to environmental literacy (*Learner Guidelines*); quality environmental education practice (*Initial Preparation Guidelines*); and quality of environmental education materials (*Materials Guidelines*).

CONCLUSION

Establishing the National Project for Excellence in Environmental Education was not without controversy. Reflecting the terminology being used by the education reform movement of the 1990s, the initiative was originally named the National Environmental Education Standards Project (Simmons1995). Using the term "standards" became problematic. Some within the environmental education community argued, and continue to argue, that it was important to use the terminology of education reform. During this time period, student achievement standards were being written by the various traditional academic disciplines (e.g., science, geography, history). There was a feeling that if environmental education was to be accepted as a legitimate field within formal education, environmental education needed to establish standards of its own. On the other hand, some within the environmental education community felt that the very notion of setting standards reflected behaviorism, elitism, and rigidity (Wals and van der Leij 1997).

Although this debate was informative, changing the name of the project was primarily in response to national education politics. As

Table 8.5. *Summary of* Non-formal Environmental Education Programs: Guidelines for Excellence (NAAEE 2004d)

Key Characteristic No. 1: Needs assessment

Effective Environmental Education Programs are designed to fill specific needs and produce tangible benefits commensurate with their costs.

1.1 Environmental Condition or Issue
1.2 Inventory of Existing Programs and Materials
1.3 Audience Needs

Key Characteristic No. 2: Organizational needs and capacities

Effective Environmental Education Programs are supportive of their parent organization's mission, purpose, and goals.

2.1 Consistent with Organizational Priorities
2.2 Organisation's Need for the Program Identified
2.3 Organization's Existing Resources Inventoried

Key Characteristic No. 3: Program scope and structure

Effective Environmental Education Programs should function within a well-defined scope and structure.

3.1 Goals and Objectives for the Program
3.2 Fit with Goals and Objectives of Environmental Education
3.3 Structure and Delivery
3.4 Partnerships and Collaboration

Key Characteristic No. 4: Program delivery resources

Effective Environmental Education Programs require careful planning to ensure that needed, well-trained staff, facilities, and support materials are available.

4.1 Assessment of Resource Needs
4.2 Quality Instructional Staff
4.3 Facilities Management
4.4 Provision of Support Materials
4.5 Emergency planning

Key Characteristic No. 5: Program quality and appropriateness

Effective Environmental Education Programs are built on a foundation of quality instructional materials and thorough planning.

5.1 Quality Instructional Materials and Techniques
5.2 Field Testing
5.3 Promotion, Marketing, and Dissemination
5.4 Sustainability

Key Characteristic No. 6: Evaluation
Effective Environmental Education Programs define and measure results in order to improve current programs, ensure accountability, and maximize the effects of future efforts.

6.1 Determination of Evaluation Strategies
6.2 Effective Evaluation Techniques and Criteria
6.3 Use of Evaluation Results

NAAEE began the development of the various documents, conversations were held with representatives of the US Department of Education. Their mandate from the United States Congress was to facilitate the development of standards in specific disciplines (i.e., science, mathematics, reading, history, civics, geography). They were not supporting the development of standards by other than these mandated disciplines. At this same time, standards developed by some of the mandated disciplines came under fire politically. For example, the history standards were vilified by the United States Senate (Chapin 1995; Nash and Dunn 1995). Since it was felt by NAAEE that developing relationships within the formal education sector was an important part of the process of developing frameworks, the project was renamed, in 1996, the National Project for Excellence in Environmental Education.

The guidelines documents were developed using a review and comment process. Public participation, of any kind, takes a concerted effort and has its pitfalls. It is costly and time consuming but, in the end, the hope is that the product benefits from these efforts. The process was well publicized in order to make participation by a wide variety of stakeholders possible. Therefore, the sheer number of responses was enormous. Each and every comment and correspondence was considered as the documents were revised. In opening the process to a wide spectrum of interests, a variety of views were tapped. Each set of guidelines benefited from this. For example, the original outline of the *Learner Guidelines* suggested seven strands. By the time the document was published, four strands had been fully developed.

During this process, individuals and organizations also began to critically analyze their own practice and the practice of environmental education. As mentioned previously, however, the development of guidelines or standards for the field of environmental education was not without some controversy. Some praised the work, while others questioned the very notion of developing guidelines. The *Canadian*

Journal of Environmental Education dedicated its second volume to this issue (see Wals and van der Leij 1997; Roth 1997; McClaren 1997), while others have worked to propose alternative models (Hart *et al.* 1999). It is hoped that this kind of examination and dialog will be ongoing even after the last set of guidelines has been published. It should also be recognized that no set of guidelines can be complete, and there are, no doubt, important characteristics missing. The *Guidelines for Excellence Series* was developed to provide a foundation on which to build systems of evaluating learning outcomes, teacher preparation and resources for environmental education that work for different people in different situations. They were developed as a tool to help inform professional judgment.

The most recent education reform movement of the 2000s in the United States has focused on standards setting and accountability within K-12 formal schooling. Academic standards for student achievement and teacher performance have been set at both the national and state levels. Accountability measures, typically in the form of state-mandated standardized testing of students at various grade levels for specific subjects (most often reading and mathematics), have been instituted widely. These standardized tests are being used to "grade" the performance of schools, school districts, and in some cases individual teachers. Teachers and teacher preparation institutions are also being held to newly developed standards, with many states requiring that teachers pass subject knowledge and skills tests before they become fully certified to teach. It should be remembered that, since environmental education is not a core discipline, it has not been included in these assessment schemes. It is difficult to predict what effect this next round of reform will have on the field of environmental education and its place in teacher education and schools.

REFERENCES

American Forest Foundation. (1997). *Project Learning Tree.* Washington, DC: American Forest Foundation.

Braus, J. (1998). *The Biodiversity Collection: A Review of Biodiversity Resources for Educators.* Washington, DC: World Wildlife Fund.

Chapin, J. R. (1995). The controversy on national standards for history. Paper presented at the Annual Meeting of the National Council for the Social Studies (Chicago, IL, November 9).

Disinger, J., and Monroe, M. (1994). *Defining Environmental Education*, An EE Toolbox Workshop Resource Manual (National Consortium for Environmental Education). Ann Arbor, MI: Regents of the University of Michigan.

Hart, P., Jickling, B., and Kool, R. (1999). Starting points: questions of quality in environmental education. *Canadian Journal of Environmental Education*, **4**, 104–124.

McClaren, M. (1997). Reflections on "alternatives to national standards in environmental education: process-based quality assessment." *Canadian Journal of Environmental Education*, **2**, 35–46.

McKeown-Ice, R. (2000). Environmental education in the United States: a survey of preservice teacher education programs. *Journal of Environmental Education*, **32**(1), 4–11.

Nash, G. B.,and Dunn, RE.(1995). National History Standards: controversy and commentary. *Social Studies Review*, **34**(2), 4–12.

National Commission on Excellence in Education. (1983). *A Nation at Risk: The Imperative for Education Reform*. Washington, DC: US Government Printing Office.

National Geographic Research and Exploration. (1994). *Geography for Life: National Geography Standards*. Washington, DC: Author.

North American Association for Environmental Education. (2004a). *Environmental Education Materials: Guidelines for Excellence*. Washington, DC: NAAEE.

(2004b). *Excellence in Environmental Education: Guidelines for Learning (K-12)*. Washington, DC: NAAEE.

(2004c). *Guidelines for the Initial Preparation of Environmental Educators*. Washington, DC: NAAEE.

(2004d). *Non-formal Environmental Education Programs: Guidelines for Excellence*. Washington, DC: NAAEE.

Roth, R. E. (1997). A critique of "alternatives to national standards for environmental education: process-based quality assessment." *Canadian Journal of Environmental Education*, **2**, 28–34.

Sanera, M. and Shaw, J. (1996). *Facts Not Fear: Teaching Children about the Environment*. Washington, DC: Regnery Publishing.

Simmons, D. (1995). *The NAAEE Standards Project*. Rock Spring, GA: NAAEE.

Simmons, D. and Vymetal-Taylor, J. (1997). *The Environmental Education Collection: A Review of Resources for Educators, Volume One*. Rock Spring, GA: NAAEE.

Simmons, D. and Vymetal-Taylor, J.

(1998a). *The Environmental Education Collection: A Review of Resources for Educators, Volume Two*. Rock Spring, GA: NAAEE.

Simmons, D. and Vymetal-Taylor, J.

(1998b). *The Environmental Education Collection: A Review of Resources for Educators,Volume Three*. Rock Spring, GA: NAAEE.

Wals, A. E. J., and van der Leij, T. (1997). Alternatives to national standards for environmental education: process-based quality assessment. *Canadian Journal of Environmental Education*, **2**, 7–27.

DAVID L. HAURY

9

Assessing the educational dimension of environmental education resources provided by non-formal groups

INTRODUCTION

In recent years there have been criticisms of environmental education in general, and of environmental education materials in particular. Sanera (1998) claims that the most widely used environmental education resources are both inaccurate and biased, and generally supportive of a "doomsday" approach to environmental issues. The Independent Commission on Environmental Education (1997) and Kwong (1995) on behalf of the Center for the Study of American Business make similar criticisms. Other non-profit organizations, such as The Consumers Union (1998) assert that: "promotional sponsored education materials blur the line between education and propaganda and lead to distorted lessons."

A shortage of funds for books and other instructional materials in schools has created a climate of receptivity and vulnerability among teachers to organizations and agencies offering free curriculum materials and media. Although schools may be underfunded when it comes to providing adequate curriculum materials, teachers have lots of help when it comes to environmental education resources for classrooms, particularly for topics related to energy industries, wood products industries, use of natural resources, or popular global issues, such as global warming, pollution, endangered species, or human population growth. Resources are offered by a broad range of non-formal education providers, such as parks, museums, and zoos, to a wide range of special interest groups and advocacy groups, to the outreach arms of major corporations and industry associations. As Smith (2000: 10) has pointed out:

> industry-sponsored materials . . . are provided for free to teachers who may lack the time and background to provide a systematic analysis and critique of them. It is not unreasonable to worry, then, that there is a very real risk that one-sided, industry-sponsored material may in some

> districts end up filling a vacuum, creating the very lack of balance that environmental education's critics decry from the opposing vantage point.

Of the environmental education resources reviewed by various groups recently (NAAEE 2000a, 2000b, 2000c; ICEE 2003), materials range from the exemplary to the factually accurate and relatively unbiased, while others present very distorted views of environmental issues, choices, and trade-offs.

The issue considered in this chapter is how to assess the potential educational value of the myriad resources being produced and disseminated to schools and teachers by non-formal groups. First, the context for considering this issue is described, followed by a four-part discussion of how the use of instructional materials relates to curriculum decisions in general, from the perspectives of the teacher, the subject matter, the learner, and the cultural milieu of schools and communities. It is suggested that the educational value of environmental education resources be assessed in terms of how the resources relate to effective instructional strategies, the aims for environmental education, the stages of student learning, and the context of the local community.

ASSESSING THE EDUCATIONAL VALUE OF RESOURCES

Educators who are committed to bringing an environmental dimension or perspective to their courses face a dilemma: though citizens express concern about the environment and a majority engage in activities to protect the environment, they seemingly lack basic knowledge about the nature and causes of environmental problems. In the United States, for example, the last five annual NEETF/Roper surveys (National Environmental Education and Training Foundation 1997, 1998, 1999, 2001, 2002) show that adults overwhelmingly support environmental education in schools and engage in activities that benefit the environment, even though their knowledge of environmental issues is weak. Given this need, educators need to be able to assess the educational value of environmental education resources, not only in terms of perspectives presented by a resource, but also within the larger context of where students obtain most of their environmental information.

Instructional resources should be reviewed and assessed with two critical realizations in mind: (a) every resource is a voice for some person, group, agency, or organization; and (b) every voice conveys an opinion and unique perspective on the topics or issues being addressed. Any book, article, essay, video, or multimedia production that addresses an environmental issue, problem, or concern reflects a point of view – the perspective of the author(s), or producers of the resource material. This is natural, and to be expected, not something to find surprising or to resist. It is an acknowledgment that must be factored into instructional decisions made by teachers and curriculum specialists. There are continua of opinions and perspectives on all matters that have a political, social, cultural, or historical context. Though perspective bias is inherent in all instructional resources that focus on such matters, underlying perspectives and biases are likely to be strongest in materials developed and produced by groups and individuals not engaged in producing textbooks or other formal instructional materials. Materials produced by non-formal groups can be expected to embody or emphasize a particular perspective or set of values that are reflective of a special interest group. The question is not whether such materials are biased or balanced in their implicit views, but how they are most productively used in a classroom.

Another factor to keep in mind is that schools are not the primary sources of environmental information for most students; that honor falls to television. In a nationwide study of young people's attitudes in the United States toward the environment, over seven out of ten students indicated that they primarily learn about the environment from television news or nature shows (Roper Starch Worldwide 1994). A slightly higher percentage indicated that school is the preferred source of information, yet television as a preferred information source increases with age. Whether or not one finds this desirable, the reality is that schools are not the primary source of information about the environment among young people, and probably never will be. Therefore, educators must carefully consider the unique contribution that schools can make, and focus efforts on that unique contribution. The key, it seems, is for schools not to compete with television as an information source, but to provide a different sort of learning experience that focuses on evaluating and using information to inform personal decisions and actions.

In any given instructional situation, someone (a teacher) is attempting to teach something (subject matter) to someone else (one or more learners) within an environment having social, physical, and cultural dimensions (a milieu). These four elements, or commonplaces (Schwab 1973): teacher, subject matter, learner, and milieu collectively comprise the educational dimension of curriculum decisions, including the selection of instructional materials. In making such decisions each of the commonplaces must be considered carefully and the relationships among them should serve as points of reflection by teachers (Posner 1985).

In the case of selecting instructional materials, then, more is required than to simply critique the materials themselves to determine "educational quality." In fact, it may be argued that an assessment of quality is secondary to the extent to which materials support the commonplaces of an educational setting. One cannot really determine the value or quality of materials in a particular educational setting without first assessing how the materials relate to the means and methods of instruction (teacher); the aims and content of instruction (subject matter); the experiential, developmental, and achievement levels of learners (student); or the priorities, values, expectations, and facilities of a school and members of the community (milieu). In short, from an educational perspective, one cannot determine the instructional quality or value of materials by examining only the materials; rather, resources must be considered within the context of each of the commonplaces of the educational setting.

The teacher: resources in context of instructional means and methods

Teachers ultimately decide what resources will be used, what is presented, how it is to be processed, and whether it is to be valued. With regard to the materials of instruction (whether textbooks, informational literature from organizations, online Web resources, or any other media type), material resources enable instruction, but do not dictate or displace a teacher's aims. Material resources are tools in the hands of effective teachers who engage students with ideas, perspectives, and choices. In this context, how does one assess the educational dimension of environmental education resources provided by non-formal groups?: by assessing the extent to which the resources enable effective instruction, both in terms of supporting effective instructional strategies, and in terms of alignment with goals and objectives of instruction.

In terms of instructional strategies, there has been much research in recent years into strategies that promote student achievement. In a summary of research-based instructional strategies for increasing student achievement (Marzano *et al.* 2001), instruction that engages students in identifying similarities and differences was noted as being particularly effective. The mental operations associated with identifying similarities and differences seem fundamental to human thought (Medin *et al.* 1995), and activities based on classifying (Ripoll 1999) or instruction using symbolic representations of similarities and differences (Lin 1996) seem particularly effective. More specifically, it has been determined that presenting students with explicit guidance in identifying similarities and differences (Chen *et al.* 1996) or asking students to independently identify similarities and differences (Mason and Sorzio 1996) enhance both students' understanding and their ability to use knowledge.

What does this have to do with environmental education materials produced and disseminated by non-formal groups? Put simply, such materials enable teachers to use one of the most effective instructional strategies available: comparisons of similarities and differences in perspectives, messages about environmental conditions and issues, values, desirable actions, and the potential long-term outcomes of local and global activities. Indeed, for such materials to be used most effectively, it is important that they not be homogeneous or neutral, but authentically reflect the various perspectives and positions of the groups or organizations that produced the materials. For instance, Smith (2000: 7) reports how one environmental educator makes a practice of "pairing materials produced by corporations and environmental advocacy groups to encourage students to examine the arguments and make up their own minds about particular topics."

Instructional materials also facilitate effective teaching when they are well aligned with the goals and objectives of instruction. Whatever subject matter is being considered, teachers must make practical decisions about the fundamental purposes of day-to-day instruction. One longstanding issue is whether instruction should focus on "content" or "process." That is, are the most important outcomes of instruction related to the learning of particular facts and concepts, or is it more important to learn skills of thinking, solving problems, making decisions, and evaluating knowledge claims? As with most issues in education, no definitive answer for all situations is possible, but it has become quite clear that, as issues become more complex and our collective knowledge expands, it is becoming increasingly important for citizens to learn how to effectively process and apply the abundance of information available to them. Within the context of environmental awareness and issues, this includes becoming aware of competing interpretations of environmental data and competing perspectives of priorities for action. Teachers having "process" goals and objectives, then, would actively seek materials that represent the broad range of opinions, interpretations, and priorities with regard to environmental issues. A diversity of views and perspectives would be considered desirable, not a threat to quality education.

Finally, many effective instructional methods engage groups of students in cooperative or collaborative learning, a social process where understanding emerges from hearing and examining the variety of ideas presented and defended by group members. Cooperative or collaborative learning activities can take many forms, from debate teams or role-playing activities to group investigations or projects. Whatever form the activities take, learning comes through dialog, expressing and reacting to ideas, and building a working consensus. Central issues in this process are "what we do with information, how we talk with each other about our values and ethics, how we work together given that we do not all advocate the same position, even when given the same information" (Cairns 2002: 85). Through such experiences, students gain more than conceptual knowledge; they gain experience in the give and take of the culture and social fabric of democratic societies. Everyone has a set of evolving values and ethics that tend to support particular perspectives on issues, so the educational need is to foster an enriched awareness of the full range of perspectives held by individuals and groups, and have students learn how to negotiate toward a functional consensus when there are widely differing views influencing decisions or actions that will affect everyone. The use of materials from non-formal groups who have perspectives that can be discerned and argued about can authenticate and enrich cooperative and collaborative activities that are used to highlight issues and problems of local or widespread import.

The subject matter: resources in the context of aims and content of instruction

Environmental education is not simply a subject to be studied, a "form" of education, nor is it simply a tool for environmental problem-solving and management (Sauvé 2002). Environmental education is, rather, an essential dimension of basic education in our common "home of life" where our primary concerns should include our community and cultural relationships with the environment. Orr (1992: 90) characterized all education as environmental education, pointing out: "by what is included or excluded, emphasized or ignored, students learn that they are part of or apart from the natural world." He further noted: "environmental issues are complex and cannot be understood through a single discipline." More recently, Davis (1998: 146) expanded this idea to say that environmental education "is interdisciplinary, mutli-disciplinary and super-disciplinary. It is about values, attitudes, ethics and actions.

It is not a subject or an 'add-on'. Nor is it an option. It is a way of thinking and a way of practice."

This multidimensional conception of environmental education cuts across the standard disciplines of schooling, and indeed, extends beyond the walls of schools. So, the challenge for formal education is to determine which aspects of this basic dimension of education to situate in schools, and to decide where best to give it form and substance within the curriculum. It is worth noting that the practice of organizing instruction around environmental themes has been shown to improve performance in standard subjects, including science, mathematics, and reading (Lieberman and Hoody 1998; Glenn 2000).

Though it is beyond the scope of this chapter to consider what aspects of environmental education to situate in schools, it is useful to consider where environmental education is located in the school curriculum when attempting to define the "subject matter" of environmental education.

Aims of environmental education

Though a latecomer among the array of recognized school subjects, environmental education as an area of study has quickly evolved around a core set of goals that have remained fairly stable over time. In the first issue of the *Journal of Environmental Education*, Stapp *et al.* (1969) stated that "environmental education is aimed at producing a citizenry that is knowledgeable concerning the biophysical environment and its associated problems, aware of how to solve these problems, and motivated to work toward their solution." In short, three familiar elements of environmental education – awareness, knowledge, and behavior – were identified early on as central goals.

On a much broader international level, the United Nations Educational, Scientific, and Cultural Organization (UNESCO) defined environmental education as:

> a process aimed at developing a world population that is aware of and concerned about the total environment and its associated problems, and which has the knowledge, attitudes, motivations, commitments and skills to work individually and collectively toward solutions of current problems and the prevention of new ones. (UNESCO/UNEP 1976: 1)

Within the Belgrade Charter we once again discern the dimensions of awareness, knowledge, and behavior, with behavior being expanded to include attitudes, motivations, commitments, and skills. The world's first Intergovernmental Conference on Environmental Education (UNESCO/UNEP 1978) led to the Tbilisi Declaration that also identified three core goals: (a) foster clear awareness of and concern about economic, social, political, and ecological interdependence in urban and rural areas; (b) provide every person with opportunities to acquire the knowledge, values, attitudes, commitment, and skills needed to protect and improve the environment; and (c) create new patterns of behavior exhibited by individuals, groups, and society as a whole toward the environment. These aims for environmental education were formulated for the purpose of promoting environmental literacy, and nurturing the development of engaged, responsible citizens prepared to recognize and actively respond to environmental issues of local or global importance. Over the years the goals of environmental education have been studied, reviewed, and adjusted, but the core dimensions have been remarkably stable. For a compilation of charters, declarations, conceptual frameworks, models, standards, and guidelines for environmental education see Thomson (2002).

Recognizing the need to align the emerging goals of the environmental education community with the standard curriculum frameworks of schools, the North American Association for Environmental Education (NAAEE), through the National Project for Excellence in Environmental Education, formulated guidelines for environmental education based on a vision of what an environmentally literate person should know and be able to do (Archie 2000; NAAEE 2000d). The guidelines are both practical and unique in recasting the broadly recognized core concepts and skills of environmental literacy into a framework that meshes well with the national curriculum standards of traditional school subjects. Indeed, care has been taken to show explicitly the linkages between the environmental education guidelines and standards in the arts, civics, economics, English, geography, history, mathematics, science, and social studies. Linkages with the science benchmarks of *Project 2061* (AAAS 1993) have also been made.

In discussing the evolution of the definition of environmental literacy, Roth (1992) presented a four-strand structure that presaged current guidelines: a knowledge strand, an affective strand, a skill strand, and a behavior strand. The *NAAEE Guidelines* (2000e) incorporate the evolving traditions of environmental literacy and action-based environmental education into four strands: questioning and analysis skills, knowledge of environmental processes and systems, skills for understanding and addressing environmental issues, and personal and civic responsibility. Each guideline focuses on a level of knowledge or skill to be exhibited by students in grades 4, 8, and 12.

There are also relevant guidelines for environmental education resources contained within the *National Science Education Standards* (NSES) (National Research Council 1996). The NSES include standards for science teaching, professional development, assessment, science content, science education programs, and education systems. Within the science content standards, one strand focuses on "Science in Personal and Social Perspectives," with special attention being given to personal and community health, population growth, natural resources, environmental quality, natural and human-induced hazards, and science and technology in local, national, and global challenges. Specific guidelines related to these topics are offered for students in grades K-4, 5–8, and 9–12. Though these are science education standards, the introduction of a social context for applying the ideas of science, and the emphasis on informed decision-making by citizens transforms these into *de facto* environmental education standards as well. Indeed, both the NAAEE *Guidelines for Learning* and the *National Science Education Standards* promote attention to making decisions, solving problems, and taking action in the context of civic responsibility. These life skills, in turn, depend on learners developing their information-seeking skills, gaining proficiency in evaluating information and information sources, and developing understanding from considering alternative or multiple perspectives.

Most non-formal groups recognize the importance of national curriculum standards to environmental education and identify explicit linkages between their materials and national standards. For instance, three of the largest environmental education projects developed by non-formal groups – Project Learning Tree, Project WET, and Project WILD – have correlated their curriculum frameworks with national and state standards (Archie 2001). Classroom teachers and workshop facilitators can use these linkages to design curricula and plan classroom instruction that is responsive to standards-based guidelines. Teachers wishing to use materials from other non-formal groups and forge linkages with national standards should consult an electronic resource – *Meeting Standards Naturally* – developed by the Environmental Education and Training Partnership (EETAP) (see http://www.eetap.org/html/meeting_standards_naturally.php).

Content of environmental education

Whether one thinks of environmental education as a separate area of study, as a broad background or dimension to all areas of study, or as a field or subject having knowledge and skill domains applicable to many other subject areas (Lowe 1998), many have expressed the importance of expanding the subject matter of environmental education beyond facts and knowledge about the natural environment. As Simmons (1998) has mentioned, environmental education needs to facilitate the thinking skills and predispositions that promote civic responsibility, enabling students to ask questions and solve problems. It is broadly believed that the essence of effective environmental education is the promotion of environmental literacy among citizens who embrace civic responsibility for the care of the environment and who are enabled to evaluate conditions, think critically about environmental issues and make informed decisions, and skillfully solve problems for the common good. As illustrated by references above to the *National Science Education Standards*, attending to literacy, social perspectives, civic responsibility, and personal decision-making is not unique to environmental education, but environmental education is unique in directing attention to using our skills, knowledge, and civic actions to solve environmental problems and sustain environmental quality. In this context, the value of educational materials provided by non-formal groups becomes clear. Informed citizens need to be aware of the diversity of activities that have an impact on our shared environments, they need to hear the diversity of perspectives about what constitutes appropriate or responsible action, they need to gain skills in critiquing messages, arguments, and calls for action, and they need practice in thinking their way through to personal decisions and actions.

Structured instructional materials have often been used to help students gain awareness of diverse perspectives and clarify their own opinions. For instance, the *Opposing Viewpoints* series included a volume on *The Environmental Crisis* (Bach and Hall 1986) that included a compilation of opposing viewpoints gleaned from magazines, journals, books, newspapers, and position statements of several organizations. The individual readings were clustered with leading questions, such as: “Is there an environmental crisis?” and “Have pollution regulations improved the environment?” In reading and discussing the various essays, students also practiced specific critical thinking skills by evaluating sources of information, separating fact from opinion, distinguishing bias from reason, understanding words in context, distinguishing primary from secondary resources, identifying stereotypes, recognizing ethnocentrism, and recognizing deceptive arguments. The value of instructional materials like this is that students are able to practice critical thinking skills on authentic readings that informed citizens would encounter in daily life. The downside of such structured materials is that the content becomes quickly dated, frozen in time, and the chosen readings must be broadly applicable to a variety of community settings. A more effective instructional strategy would couple the use of critical thinking skills with the reading of materials that present diverse perspectives on issues of current interest and are of local relevance.

Resources from non-formal agencies – whether they represent environmentalists, conservationists, or advocates for commercial exploitation of natural resources – can be invaluable tools for teachers who focus on critiquing evidence and arguments, developing awareness of multiple perspectives, raising awareness of environmental issues and problems, or nurturing personal and collective decision-making and action. From an educational perspective, the value of such materials increases dramatically if they are timely and can be linked to issues of local relevance. It has been stated by many environmental education advocates that teachers ought not to be strong advocates of particular courses of environmental action, but they should be enabling their students to make responsible decisions and take informed actions that reflect their understanding and values.

In the domain of subject matter, resources from non-formal groups broaden the knowledge base and provide an authentic continuum of voices and messages for students to critically examine. Hungerford *et al.* (2003) provide an example of how such materials can be incorporated into skill development. Their highly acclaimed program was designed to help students learn how to investigate and evaluate science-related social issues, and how to locate, analyze, and use information sources, including a variety of non-formal groups among government and private agencies. In this example, students themselves seek materials from non-formal groups and use the materials to gain authentic experience in critically examining information sources and evaluating their content.

The learner: resources in the context of experiential and developmental levels

A focus on the learner implies some understanding of learning and one important fact that we know for sure is that students are not empty vessels into which teachers pour factual knowledge. Not only do facts not equal knowledge, it is the learner who must construct knowledge from his or her experiences and the information at hand. Further, "humans engage in thinking, feeling, and acting, and these combine to form the meaning of experience" (Novak 1998: 9). Therefore, education must focus on three forms of learning: acquisition of knowledge (cognitive learning), changes in emotions or feelings (affective learning), and gains in physical actions or performance (psychomotor learning). In short, learning is an active process that involves constructing meaning from experiences, and experiences include dimensions of thinking, feeling, and acting. This is, admittedly, an oversimplification of the complexities of learning, but it is sufficient to make a critical point: meaningful learning involves much more than rote learning of facts; it results from the union of actions, feelings, and conscious thought (Novak 1998: 31). By extension, meaningful learning in environmental education would result from engaging individuals or groups of students in activities or in making decisions anchored to issues or topics of local interest or relevance that require critical thinking or application of knowledge.

When considering the process of learning, it is important also to acknowledge that the development of learners progresses in predictable phases over time, in terms both of cognitive development (Furth 1970; Labinowicz 1980; Johnson 1982;) and moral development (Kohlberg and Hersch 1977). The implications from both cognitive and moral development are that: (a) there is a progression with regard to what can be expected of learners across the K-12 spectrum, and (b) sometime during middle school or after, most learners become capable of abstract reasoning and exhibiting sensitivity to the values of others. The implication for environmental education classrooms is that a focus on citizen action skills and the development of a personal sense of moral ethics is probably not appropriate prior to middle school or upper middle school grades. Therefore, the use of materials from non-formal groups for the purpose of examining competing perspectives and clarifying one's own values is most appropriate from middle school grades on. Engleson and Yockers (1994) provide a brief summary of research findings and implications for environmental education at various grade levels.

In considering learners, it is also important to realize that there are multiple dimensions of intelligence (Gardner 1983) and a variety of learning styles or preferences (McCarthy 1987; Dunn 2003). Though there are uncertainties about the optimal means of nurturing the various domains of intelligence, and there are competing theories regarding the nature of learning styles, two things are clear: (a) there are various forms of "knowing" associated with the various forms of intelligence, and (b) different learners have different preferences in modes of learning. A very general implication of these findings is that learning in general is likely to increase when teachers engage multiple domains of intelligence using a variety of instructional modes. The primary implication for the materials of instruction is that variety in terms of perspectives and modes of engaging students with multiple perspectives will likely increase the level of learning and understanding.

Finally, it is obvious from looking into classrooms that learners are diverse in many ways, in terms of ethnicity, beliefs, religious heritage, socioeconomic status, life experiences, and cultural background. Instruction and the tools of instruction must both reflect this diversity and exhibit respect for diversity.

The milieu: resources in the context of schools and communities

There are many factors influencing schools, teaching, and programs that coalesce to form the context for teaching and learning. Here I will focus on only two factors: situated learning and life-long learning.

Though it is important to have national education standards that ensure that all students have the same educational foundation on which to stand, it is also important to acknowledge that everyone learns within the context of a local community. Schools are agencies of local communities, and each community has its own environmental issues, resources, and limitations. The environmental cliché "think globally, act locally" is both environmentally and educationally sound. All global environmental issues have local ramifications that can be directly experienced, and school topics become increasingly authenticated and relevant as they are more closely linked to local conditions and concerns. An advantage of using a selection of materials produced by nonformal groups is the increased possibility of tailoring the selection of materials to local conditions, priorities, and issues. Such a practice is both educationally sound and of long-term value to the community as school learners develop to become informed citizens.

Secondly, in a world where the one constant is change, and where the expansion of knowledge is vastly outstripping anyone's ability to stay informed in all fields of relevance to informed citizenship, gaining proficiency in the skills of self-directed, life-long learning is crucially important. Indeed, some have claimed: "the central purpose of education is to empower learners to take charge of their own meaning-making" (Novak 1998: 9). It could be argued from an educational point of view that there is particular merit in having students learn effective information-seeking behaviors in addition to learning how to detect the bias inherent in advocacy materials. Since formal education cannot provide learners with all the knowledge content that they will need throughout their lives, it is important that they learn the skills both of finding information that informs decisions and actions, and of critical thinking that will serve them in evaluating information sources, enabling them to discern opinion and bias in writing and presentations.

CONCLUSION

In considering the merits of using educational materials produced by non-formal groups, education can be broadly conceived as comprising four commonplaces: teachers, learners, subject matter, and a context or milieu. Each of these commonplaces has implications for both the choice of materials used, and the methods associated with their use. With regard to the instructional methods used by teachers, one of the most effective instructional methods is to examine similarities and differences. In recent years, instruction has also increasingly focused on processes of learning and thinking rather than on a static body of content knowledge, and instruction has become decidedly more social as cooperative and collaborative instructional strategies compete with more traditional methods where students work alone. The use of educational materials produced by non-formal groups fit well with these instructional strategies, enabling students to examine voices and messages for similarities and differences, gain practice in the skills of evaluating sources of information, and working in teams to locate, critique, and apply knowledge gleaned from such material.

With regard to subject matter, environmental education takes a variety of forms from school to school, but over time a consensus has formed regarding the primary goals of environmental education. Though there are important concepts to be learned in the domain of environmental education, much more emphasis has been placed on critical thinking, personal decision-making, and civic action. Here, too, the opportunity to critically examine the multiple perspectives embedded in materials produced by non-formal groups has significant educational value. If adult citizens are to make informed decisions and become meaningfully engaged in civic actions, they must first learn the skills associated with becoming informed, becoming aware of multiple perspectives, and evaluating the rationales associated with possible actions. Exposure to a variety of voices and messages conveyed through materials produced by non-formal groups allows young citizens to gain awareness and practice skills associated with testing ideas and establishing one's own environmental ethic.

The primary lesson we learn from studying the nature of learners is three-dimensional: (a) learners construct their own knowledge from their unique sets of experiences; (b) learners of any given grade or age are at various levels of cognitive and moral development; and (c) learners display a variety of intelligences and learning styles. In a word, all learners differ in a variety of fundamental ways, and their readiness to learn abstract content or higher-level skills depends on their developmental pathways. There are two overarching implications for curriculum and instruction: (a) there is no one best instructional approach that is most effective with all learners; and (b) instruction at all levels is most effective when it is developmentally appropriate. The implications for the use of materials by non-formal groups are that: (a) learning will be most effective for all learners in a group when there is opportunity to interact with materials in a variety of ways, and (b) using materials to promote personal decision-making and consideration of civic action is not likely to be effective prior to the middle school or upper middle school grades.

Finally, the educational milieu of schools has both a local and a global dimension. At the local level, schools are situated in communities that have particular environmental conditions, issues, and concerns. The more closely that selection of materials can reflect the local environmental milieu, the better. This increases both the authenticity and the relevance of instruction. In a global sense, we are in an age where mastery of skills associated with life-long learning is important to all learners, regardless of their local context. In order to be meaningfully engaged in ongoing dialog and decision-making related to environmental issues, one must be able to continually engage in self-directed learning as conditions change and our knowledge base expands. Here, too, materials produced by non-formal groups offer a distinct advantage; teacher selection of timely and locally relevant materials enables learners to practice skills associated with life-long learning with materials, ideas, and arguments that are current and immediately relevant, not contrived .

In short, if developmental readiness of learners is taken into account, and provision is made for the continua of differences among learners, materials produced by non-formal groups can be used very effectively in environmental education to support the most effective instructional strategies, the current goals of instruction, and the context of instruction. From this educational perspective, how a teacher uses such materials to engage learners is more critical than the inherent quality of the materials. Even so, there are robust guidelines for teachers to follow in selecting materials (refer to Appendices 9.1, 9.2, 9.3 and 9.4).

REFERENCES

American Association for the Advancement of Science. (1993). *Project 2061: Benchmarks for Science Literacy.* New York, NY: Oxford University Press. [Online] URL: http://www.project2061.org/tools/benchol/bolframe.htm.

Archie, M. (2000). *Excellence in Environmental Education: Guidelines for Learning (K-12).* Washington, DC: North American Association for Environmental Education. [Online] URL: http://naaee.org/npeee/learner_guidelines.php.

(2001). *EETAP Celebrates: Five Years of Advancing Education and Environmental Literacy... and Next Steps.* Washington, DC: North American Association for Environmental Education.

Bach, J. S. and Hall, L., eds. (1986). *The Environmental Crisis: Opposing Viewpoints.* St. Paul, MN: Greenhaven Press.

Branch, R. M., Kim, D., and Koenecke, L. (1999). Evaluating online educational materials for use in instruction. Syracuse, NY: ERIC Clearinghouse on Information & Technology. [Online] URL: http://ericit.org/digests/EDO-IR-1999-07.shtml.

Cairns, K. (2002). The legitimate role of advocacy in environmental education: how does it differ from coercion? *Ethics in Science and Environmental Politics,* 82–87.

Chen, Z., Yanowitz, K. L., and Daehler, M. W. (1996). Constraints on accessing abstract source information: instantiation of principles facilitates children's analogical transfer. *Journal of Educational Psychology,* **87**(3), 445–454.

Consumers Union, The. (1998). How great a problem? In *Captive Kids: A Report on Commercial Pressures on Kids at School.* [Online] URL: http://www.consumersunion.org/other/captivekids/problem.htm

Davis, J. (1998). Young children, environmental education and the future. In *Education and the Environment,* ed. N. Graves. London: World Education Fellowship, pp. 141–154.

Dunn, R. (2003). The Dunn and Dunn learning-style model and its theoretical cornerstone. In *Synthesis of the Dunn and Dunn Learning-Style Model Research: Who, What, When, Where, and So What?,* ed. R. Dunn and S. A. Griggs. New York, NY: St. John's University's Center for the Study of Learning and Teaching Styles, pp. 1–6.

Engleson, D. C. and Yockers, D. H. (1994). *A Guide to Curriculum Planning in Environmental Education.* Madison, WI: Wisconsin Department of Public Instruction.

Furth, H. J. (1970). *Piaget for Teachers.* Englewood Cliffs, NJ: Prentice-Hall.

Gardner, H. (1983). *Frames of Mind: The Theory of Multiple Intelligences.* New York, NY: Basic Books.

Glenn, J. L. (2000). *Environment-based Education: Creating High Performance Schools and Students.* Washington, DC: The National Environmental Education and Training Foundation.

Hungerford, H. R., Volk, T. L., Ramsey, J. M., Litherland, R. A., and Peyton, R. B. (2003). *Investigating and Evaluating Environmental Issues and Actions: Skill Development Program*. Champaign, IL: Stipes Publishing.

Independent Commission on Environmental Education. (1997). *Are We Building Environmental Literacy?* Washington, DC: George C. Marshall Institute. [Online] URL: http://www.marshall.org/article.php?id=10.

(2003). *Environmental Education Materials*. [Online] URL: http://www.enviroliteracy.org/article.php/286.html.

Johnson, V. R. (1982). Myelin and maturation: a fresh look at Piaget. *The Science Teacher*, **49**(3), 41–44.

Kavassalis, C., Symington, D., and Barwell, R. (2002). *Environmental Education on the Internet*. [Online] URL: http://www.biomuncie.org/Environmental%20 Education.htm (cited February 23, 2003).

Kohlberg, L. and Hersch, R. H. (1977). Moral development: a review of the theory. *Theory into Practice*, **16**(2), 53–59.

Kwong, Jo. (1995). *Environmental Education: Getting Beyond Advocacy*. Center for the Study of American Business, Contemporary Issues Series 76.

Labinowicz, E., ed. (1980). *The Piaget Primer: Thinking-Learning-Teaching*. Menlo Park, CA: Addison-Wesley.

Lieberman, G. A. and Hoody, L. L. (1998). *Closing the Achievement Gap: Using the Environment as an Integrating Context for Learning*. San Diego, CA: State Education and Environment Roundtable.

Lin, H. (1996). The effectiveness of teaching science with pictorial analogies. *Research in Science Education*, **26**(4), 495–511.

Lowe, I. (1998). Environmental education: the key to a sustainable future. In *Education and the environment*, ed. N. Graves. London: World Education Fellowship, pp. 95–104.

Marzano, R. J., Pickering, D. J., and Pollock, J. E. (2001). *Classroom Instruction that Works: Research-based Strategies for Increasing Student Achievement*. Alexandria, VA: Association for Supervision and Curriculum Development.

Mason, L. and Sorzio, P. (1996). Analogical reasoning in restructuring scientific knowledge. *European Journal of Psychology of Education*, **11**(1), 3–23.

McCarthy, B. (1987). *The 4MAT system: Teaching to Learning Styles with Right/Left/Mode Techniques*. Barrington, IL: Excel.

Medin, D., Goldstone, R. L., and Markman, A. B. (1995). Comparison and choice: relations between similarity processes and decision processes. *Psychonomic Bulletin & Review*, **2**(1), 1–19.

National Environmental Education and Training Foundation. (1997). *The National Report Card on Environmental Knowledge, Attitudes and Behavior. The Sixth Annual Survey of Adult Americans*. Washington, DC: NEETF.

(1998). *Environmental Myths in America: An Average American View. The 1998 National Report Card: Environmental Knowledge, Attitudes, and Behavior. The Seventh Annual Survey of Adult Americans*. Washington, DC: NEETF.

(1999). *Environmental Readiness for the 21stCentury. 1999 NEETF/Roper Report Card*. Washington, DC: NEETF.

(2001). *Lessons from the Environment. The Ninth Annual National Report Card on Environmental Attitudes, Knowledge and Behaviors*. Washington, DC: NEETF.

(2002). *American's Low "Energy IQ": A Risk to Our Energy Future. Why America Needs a Refresher Course on Energy. The Tenth Annual National Report Card: Energy Knowledge, Attitudes, and Behavior*. Washington, DC: NEETF.

National Research Council. (1996). *National Science Education Standards*. Washington, DC: National Academy Press [Online] URL: http://www.nap.edu/readingroom/books/nses/html/.

North American Association for Environmental Education. (2000a). *The Environmental Education Collection: A Review of Resources for Educators, Volume One*. Washington, DC: NAAEE.

(2000b). *The Environmental Education Collection: A Review of Resources for Educators, Volume Two*. Washington, DC: NAAEE.

(2000c). *The Environmental Education Collection: A Review of Resources for Educators, Volume Three*. Washington, DC: NAAEE.

(2000d). *Excellence in Environmental Education: Guidelines for Learning (K-12): Executive Summary & Self-assessment Tool*. Washington, DC: NAAEE.

(2000e). *Environmental Education Materials: Guidelines for Excellence*. Washington, DC: NAAEE.

(2000f). *Environmental Education Materials: Guidelines for Excellence Workbook: Bridging Theory and Practice*. Washington, DC: NAAEE.

Novak, J. D. (1998). *Learning, Creating, and Using Knowledge*. Mahwah, NJ: Lawrence Erlbaum.

Orr, D. W. (1992). *Ecological Literacy: Education and the Transition to a Postmodern World*. Albany, NY: State University of New York Press.

Posner, G. J. (1985). *Field Experience: A Guide to Reflective Teaching*. New York, NY: Longman.

Ripoll, T. (1999). Why this made me think of that. *Thinking and Reasoning*, **4**(1), 15–43.

Romanello, S. (1998, updated 2003). *Information Sources for Environmental Education*. Columbus, OH: ERIC Clearinghouse for Science, Mathematics, and Environmental Education. ERIC Digest. [Online] URL: http://www.ericse.org/digests/dse98-10.html.

Roper Starch Worldwide. (1994). *Environmental Attitudes and Behaviors of American Youth with an Emphasis on Youth from Disadvantaged Areas*. Washington, DC: National Environmental Education and Training Foundation. [Online] URL: http://eelink.net/ROPER/TOC.html.

Roth, C. (1992). *Environmental Literacy: Its Roots, Evolution, and Directions in the 1990s*. Columbus, OH: ERIC Clearinghouse for Science, Mathematics, and Environmental Education.

Sanera, M. (1998). Environmental education: promise and performance. *Canadian Journal of Environmental Education*, **3**, 9–26. [Online] URL: http://www.edu.uleth.ca/ictrd/cjee/volume_3/Sanera.html.

Sauvé, L. (2002). Environmental education: possibilities and constraints. *Connect*, **27**(1–2), 1–4.

Schwab, J. J. (1973). The practical 3: translation into curriculum. *School Review*, **81**(4), 501–522.

Simmons, D. (1998). Reflections on "Environmental education: promise and performance." *Canadian Journal of Environmental Education*, **3**, 41–47.

Smith, G. A. (2000). *Defusing Environmental Education: An Evaluation of the Critique of the Environmental Education Movement* (Education Policy Project Report CERAI-00-11). Milwaukee, WI: Center for Education Research, Analysis, and Innovation, University of Wisconsin-Milwaukee. [Online] URL: http://www.asu.edu/educ/epsl/EPRU/documents/cerai-00-11.htm

Stapp, W. B., Bennett, D., Bryan, W., Fulton, J., MacGregor, J., Nowak, P., Swan, J., Wall, R., and Havlick, S. (1969). The concept of environmental education. *Journal of Environmental Education*, **1**(1), 30–31.

Thomson, G. (2002). *What Is Good Environmental Education? A Draft Backgrounder for Practitioners*. Canadian Parks and Wilderness Society. [Online] URL: http://www.cpawscalgary.org/education/network-environmental-education/whatisgoodee.pdf (retrieved October 25, 2003).

UNESCO/UNEP. (1976). The Belgrade Charter. *Connect: UNESCO-UNEP Environmental Education Newsletter*, **1**(1), 1–2.
(1978). The Tbilisi Declaration. *Connect: UNESCO-UNEP Environmental Education Newsletter*, **3**(1), 1–8.
Wyatt, V. (2003). Choosing and using good science books. *Teaching K-8*, **33**(5), 52–53.

APPENDIX 9.1. ASSESSMENT GUIDELINES

NAAEE Guidelines for materials

In addition to developing guidelines for learning in environmental education, the NAAEE has developed guidelines for selecting, evaluating, or developing environmental education materials (NAAEE 2000e). The guidelines are clustered around six key characteristics of quality environmental education materials: fairness and accuracy, depth, emphasis on skill-building, action orientation, instructional soundness, and usability. A workbook (NAAEE 2000f) has been developed to assist educators with application of the guidelines (available at: http://www.naaee.org/npeee/).

Resource assessment matrix

Given the highly contextual nature of decisions about educational resources, it is advantageous for teachers, committees, or schools to develop their own assessment matrices. An individual teacher will not likely have time to use the full range of NAAEE guidelines every time to assess a teaching resource. A more effective approach would be to have a standard set of criteria or questions to ask while reviewing potential resources (e.g., Appendix 9.2). Or, teachers may wish to develop a more extensive set of criteria (e.g., Appendix 9.3), such as those developed by Kavassalis *et al.* (2002) at the Ontario Institute for Studies in Education. This set of criteria reflects the dimensions and priorities of environmental education described in the Belgrade Charter (UNESCO/UNEP 1976).

The special case of online resources

The World Wide Web has transformed communication and greatly expanded access to environmental and educational resources. Anyone, anywhere in the world can distribute background information, reports, opinions, research findings, and any other form of communication to whoever can locate the resource and wishes to examine it. The traditional information filters associated with professional journals and the editorial control of publishing houses are removed. For a large proportion of Web-based resources, there is no editor, reviewer, or review mechanism to ensure or indicate credibility, quality, accuracy, or timeliness of resources offered. As the democratization of information increases through free, global, electronic access to information sources, it will become increasingly important for citizens both to employ effective search strategies and to evaluate the credibility and application of resources.

EE Link, a service of the NAAEE, provides a listing of directories and databases of environmental education curricula and related classroom materials at: http://eelink.net/classroomresources-directories.html. Each entry includes a brief description of the resource and an indication of ratings provided by users.

A wealth of other professional and classroom resources is also available through EE Link. Romanello (2003) has also provided a listing of credible information sources for environmental education that can be accessed at: http://www.ericse.org/digests/dse98-10.html. Other portals to environmental information and educational resources include the US Environmental Protection Agency Environmental Education Center directory of educational resources (http://www.epa.gov/enviroed/resources.html) and Exploring the Environment (http://www.cotf.edu/ete/).

Branch, Kim, and Koenecke (1999) also provide general guidelines for evaluating the suitability of using online educational materials for instructional purposes. In addition to providing a checklist for evaluating online materials, the authors provide a listing of other online tools for evaluating the educational merits of online resources. Kavassalis, Symington, and Barwell (2002) have also developed a set of criteria to differentiate environmental education from environmental advocacy. Among other resources provided is a "Decision Tree" that teachers can use to evaluate the quality and potential bias of online environmental resources. The site provides an extensive listing of references and resources regarding environmental education, advocacy and education, and environmental directories.

APPENDIX 9.2. SEVEN CORE QUESTIONS TO ASK ABOUT POTENTIAL RESOURCES FOR YOUNG STUDENTS

1. ***Does the resource speak to children?*** Beyond having accurate and relevant information, resources to be used directly by young students should relate directly to their world of experiences; there should be obvious connections to their everyday lives in terms that are familiar to them.
2. ***Does the resource speak to all children?*** Do the materials represent or respond to the diversity of children in classrooms in terms of ethnicities, heritage, experiential backgrounds, or worldviews? Does the resource appeal to girls as well as boys?
3. ***Is there a wide range of learning possibilities?*** Young people as well as older students have a wide range of interests and learning style preferences. Do resources provide learning options, a variety of direct experiences, a rich combination of textual information and visuals, and high interest elements such as games, puzzles, or stories?
4. ***Do the learning activities emphasize critical thinking, inquiry, problem-solving, or use of information over memorization of facts?*** Life-long learners must begin to recognize early on that facts can always be located, but the construction of knowledge, using information to make decisions, and solving problems requires the application of critical thinking skills.
5. ***Are the learning activities safe and simple?*** The educational value of activities should not be degraded by unduly complex or risky procedures, or procedures that are not easily adapted to the constraints of time and materials found in typical classrooms. Equipment and supply needs should include items that are typically found in classrooms and homes, and instructions should be clear to average readers. There should also be notices where adult help is needed.

6. ***Can you, the teacher, understand the content?*** You should not need advanced studies in science, economics, or other disciplines to understand the principles and concepts in environmental education materials intended for use by young students. If you as an adult do not understand the ideas being presented, young students certainly will not understand them either. It is expected that you will need to help them in their understanding, but the meaning of resources intended for young students should be readily apparent and clear to informed adults without advanced degrees in particular content areas.
7. ***Will using the resource be fun for students, or provoke interest or wonder among students?*** Is humor used to good advantage and are there attention-grabbing illustrations? Are there interesting or bizarre facts? We live in an age when the youngest students have encountered information being packaged in very appealing ways, so resources are more likely to be engaging if attention is given to presentations that invite attention. Appealing materials may not be of high substantive quality, but unappealing, high-quality materials are less likely to engage students in learning.

Source: These questions have been enhanced by ideas garnered from Wyatt (2003).

APPENDIX 9.3. EXAMPLE CRITERIA FOR ENVIRONMENTAL EDUCATION RESOURCES

A. Resources should provide **knowledge** about the environment and environmental issues. As such, resources should:

- Develop awareness and understanding of environmental concepts.
- Provide balanced presentation of differing viewpoints and theories about environmental issues.
- Provide fair and accurate information.
- Provide well-documented factual information.
- Provide information and data drawn from current and respected sources.
- Provide the identification and affiliation of the author.
- Provide the science that underlies the environmental issue or topic.
- Reflect the complexity of issues providing depth, and scale.
- Draw attention to different scales: local, national, and global.
- Help learners recognize the symptoms and causes of environmental problems including socioeconomic and political factors.
- Place current or potential environmental challenges in historical perspective.

B. Resources should provide *skills* that will enable users to learn about the environment and to address environmental issues. As such, resources should:

- Suggest a rational, analytic approach to problems.
- Provide suggestions for action, but not advocate a singular course of action.
- Support skill-building in areas such as: data collection, inquiry, critical thinking, and decision-making.

C. Resources should promote positive *attitudes* toward the environment and environmental issues. As such, resources should:

- Emphasize the interconnection between people and the environment and the need to manage and protect the environment.
- Motivate and empower people to act on their own conclusions about what should be done to resolve environmental problems.
- Promote civic responsibility and the value and necessity of local, national, and international cooperation.

Source: Kavassalis, *et al* (2002).

APPENDIX 9.4. OTHER SOURCES OF SUPPORT FOR BUSY TEACHERS

North American Association for Environmental Education (NAAEE)

NAAEE has developed two reviews of resources: *Environmental Education Collection: A Review of Resources for Educators* and *The Biodiversity Collection* (available at :http://www.naaee.org/npeee/), based on the NAAEE *Guidelines for Materials*. The reviews illustrate how the *Guidelines for Excellence* can be used to assess materials, identify the strengths and weaknesses of specific materials, and highlight topics covered, grade levels, and the school subjects emphasized for each resource. NAAEE also provides an online directory of directories for classroom resources, including many produced by non-formal groups (http://eelink.net/classroomresources-directories.html).

National Educational and Environmental Partnership (NEEP)

NEEP is a private, non-profit coalition of environmental education organizations; national associations for science, social studies, mathematics, and geography; teachers, school administrators, and school boards; non-formal education organizations; the business community; and governmental agencies. The partnership has developed a guide entitled, *Environmental Education and Educational Achievement: Promising Programs and Resources* (NEEP 2002), that provides sources for resources and sample curriculum materials (http://www.neetf.org/Education/reports.shtm). A related guide, *Environmental Education: Resources at a Glance* (http://www.neetf.org/Education/ata-glance.pdf), includes the same listing of resources, with highlights of the academic benefits of environmental education. In 1995, NEEP also conducted a survey of state-level environmental education programs in the United States, seeking answers to questions about communication and coordination between formal and non-formal environmental education efforts in each state (http://www.uwsp.edu/cnr/neeap/statusofee/outdoor.htm).

Environmental Education & Training Partnership (EETAP)

EETAP is the national training program for the US Environmental Protection Agency's Office of Environmental Education. In support of its training mission, the project provides resources (http://eetap.org/resources.html) to teachers, non-formal educators, state and local education agencies, natural resource officers, and other non-profit organizations. Current resources include: *Environmental Education Information Providers Directory* (http://www.ag.ohio-state.edu/~eetap/pdf/infopro.pdf) and *Finding Resources on the Internet: A Trainer's Module for Environmental Education* (http://www.ag.ohio-state.edu/~eetap/pdf/trainers.pdf).

Educational Resources Information Center (ERIC)

The ERIC system, funded by the US Department of Education, has developed the world's largest education-related bibliographic database (http://www.eric.ed.gov/searchdb/index.html). EETAP provides a useful guide, the *Thesaurus of Environmental Education Terms for Use in Electronic Database Searching* (http://www.ag.ohio-state.edu/~eetap/pdf/eeterms.pdf), to searching the ERIC database for environmental education materials. There are also a number of ERIC Digests focusing on environmental education resources, including the following: *Information Sources for Environmental Education, Choosing Instructional Materials for Environmental Education, Teaching About Ecosystems* and *Environmental Education Resources on the World Wide Web* (http://www.ericfacility.net/ericdigests/index/).

US Environmental Protection Agency (EPA)

The EPA provides access to an on-line catalog of environmental education materials and an index of resources by environmental topic (http://www.epa.gov/enviroed/resources.html). The site also provides links to other environmental education Web sites supported or developed by the US EPA.

STEIN DANKERT KOLSTØ

10

Assessing the science dimension of environmental issues through environmental education

INTRODUCTION

It is important that environmental education in schools and other institutions offers learners relevant tools to investigate and evaluate the science involved in debates on environmental issues. Environmental issues have social dimensions as they involve the making of personal, institutional, or cultural decisions. By their nature, however, they also have a science dimension. The science involved in some environmental issues still is often debated within the scientific community. This situation constitutes a challenge for those who want to base their evaluation, viewpoint, and actions on evidence and facts. School science traditionally focuses on established non-controversial "core concepts" of science, leaving the learners with little experience in dealing with real, or perceived, disagreement between experts, or the provisional nature of science.

The perspective in this chapter is that the aim of environmental education should be based on "thoughtful decision-making." The term "thoughtful," which is adopted from Aikenhead (1985), is used to indicate that there is not one single way to make valid or rational decisions. The hope is that citizens will base their decisions and actions on evaluations of information and evidence in interaction with their personal values. Thus the aim is not to convince the learners that one set of values and actions is correct, but to challenge the learners' views and values in order to develop their own thinking.

To arrive at a thoughtful decision a citizen needs to be competent in several ways. The citizen needs to *investigate the environmental issue*, in order to gain a deeper understanding of the issue and the controversies involved. Based on the outcome of this investigation the citizen needs to *evaluate the information* and the scenarios involved with regard to the relevance, credibility, and importance related to his/her values. In addition s/he needs to be able to *engage in debates* on the issues in order to express views, test out arguments and influence the views of antagonists and others.

To provide a foundation for this discussion, the first section of this chapter will describe *what* is to be examined; i.e., the science and complexity of environmental issues. The second section will examine our current knowledge of *how* students investigate and assess science-related information. The third section will present a framework for the development of knowledge and tools to investigate and evaluate the science of environmental issues. The framework is based on how knowledge about science can inform investigation and evaluation. The final section will discuss attitudes and skills required for thoughtful decision-making along with example teaching approaches to develop those attitudes and skills.

WHAT IS TO BE EXAMINED?

We need to have some views on what characterizes environmental issues in order to discuss what competencies students must develop to examine them in depth. A citizen decision-maker is confronted with two main questions. First, there is the ethical, personal, or social question related to what actions to take. Secondly, what is the scientific evidence presented in an environmental issue? The debate on climate change illustrates this. One focus of the debate is on what actions to take, or not to take. This is an ethical, economic, and political question. Another focus of the debate is what the evidence is for climate change and whether this change is, to some extent, caused by man.

Decision-making on environmental issues is further complicated by the fact that citizens often find contrasting expert answers to the scientific question. This situation is also evident in the climate change debate. The UN's Intergovernmental Panel on Climate Change (IPCC) has stated that: "[t] here is new and stronger evidence that most of the warming observed over the last 50 years is attributable to human activities" (IPCC 2001: 10). However, not all scientists agree with the conclusion of the climate panel. There are several solar physicists who claim that the variations in sun spot activity provide a plausible explanation for changes in the Earth's climate. Fligge *et al.* (1999: 46) state that: "[t]he length of the sunspot cycle . . . correlates well with indicators of terrestrial climate," suggesting that human activity is not the main cause of climate change. The presence of alternative interpretations is a challenge for citizens trying to form an opinion on an environmental issue.

Expert disagreement is sometimes evident as well in the information reaching the general public through the media. For example, a senior scientist working at the Norwegian Institute for Air Research stated in a newspaper interview that: "the probability that human activity influences the climate in a negative direction is very great," and "research on climate in recent years has documented a negative development and clear signs of risk" (Kaarbo 1999: 6, translation by this author). In a response to this interview a researcher at the Center for Energy and Environment Studies at the Norwegian School of Management, and a research scientist at the Institute of Theoretical Astrophysics at the University of Oslo, commented in a letter to the editor (Brekke and Anker-Nilssen 1999: 18, translation by this author): "These are overstated claims. We say that such claims are not at all documented. And besides this, the assertion that it is caused by human action is even less evidential." As the quotes indicate, findings at the frontier on science are rarely unanimous.

However, many environmental issues involve science that is uncontroversial within the scientific community; e.g., the science of cloud formation in climate models. Therefore, several scholars have suggested a distinction between two kinds of science. One kind, denoted as "core science" (Cole 1992), is characterized by a consensus within the scientific community. This is science where the scientific debates at the initial stages of the research have settled, and which now occurs in textbooks. This is the science emphasized in schools and consequently the science most familiar to us. The law of energy conservation, the hydrologic cycle, and the genetic explanation of inheritance are cases in point. The other kind, denoted as "frontier science" (Cole 1992), is science in the process of being researched. At this stage, hypotheses are being developed and tested, and results from research are presented to colleagues for peer review. Through this process of presentation, publication, and further argumentation, some concepts or explanations are becoming supported by a consensus within the relevant scientific community. Such consensus is believed to reflect the community's judgment of agreement between concepts and empirical data.

A consequence of distinguishing between the two kinds of science is an increased understanding of the legitimacy of disagreement and debate among experts. Another consequence is an increased understanding that the reliability of scientific knowledge depends on its ability to withstand criticism based on scientific norms and the strength of the consensus that supports it (Bingle and Gaskell 1994). This level of understanding makes citizens' investigation and evaluation of scientific knowledge claims rather difficult.

Results from the frontier of science are often used in environmental issues before a consensus has been reached within the scientific community. This phenomenon is due to the availability of results from frontier science combined with the urgent character of many issues. Science may take years or decades to reach consensus on scientific issues, but in most environmental controversies a decision has to be made as soon as possible, and not making any decision also implies an action. Debate among scientists, however, does not prevent citizens from judging the research and the evidence, assessing whether "one hypothesis is more probably correct or more deserving of trust or more likely than another" (Barnes 1987: 563). It again shows the importance of citizens' ability to assess the credibility of scientific claims and the relevance of teaching this in environmental education.

HOW DO STUDENTS ASSESS SCIENCE-RELATED INFORMATION?

How do students assess the complexity of environmental controversies and the science dimensions involved in environmental issues? Or, more specifically, what kind of knowledge and understanding might be relevant for students' investigation and assessment of the science involved in such issues?

A British study of 16-year-old students' arguments and views on issues related to biotechnology (Lewis *et al.* 1997) concluded that most of the students were able to consider issues in meaningful ways, even if the students' understanding of several basic concepts in biotechnology was found wanting. There are also indications that their arguments and decision-making tend to focus on ethical aspects, possible risks, and the credibility of information rather than on scientific concepts. For example, in a study of 15-year-old students debating different aspects of gene technology (Kolstø 2003a) the students appear to make use of scientific concepts. However, closer inspection of their arguments showed that it was hard to tell whether or not they understood the concepts they were using. Moreover, scientific concepts were only used in sentences describing either a technological possibility (e.g., "I think the possibility of developing genetically modified crops makes it important to inform the consumers if such crops are used in a product"), or sentences stating an issue (e.g., "Should it be legal to import genetically modified food?"). Thus it seemed as if these students used the scientific concepts not to assess areas of scientific knowledge, but to assess ethical issues and possible consequences of different actions based on technological possibilities. The students did not find it necessary to ensure that they had a common understanding of the underlying science, as long as they were able to communicate about the ethical aspects of the issues.

In another study, science education student teachers were asked to assess an article of their own choice related to a socioscientific issue (Kolstø 2003b). Analysis of their evaluations found that they seldom used scientific criteria related to the quality of methodology, theoretical consistency, or consistency in argumentation. Instead, they criticized the articles for lacking information on methodology and references. This raises the question of whether the use of scientific criteria is too demanding for most lay people. It takes effort to get access to scientific standards, and these might vary from one field of research to another (see e.g., Norris 1995).

Based on several case studies on the relation between expert knowledge and local knowledge in selected socioscientific issues, Irwin and Wynne (1996) state that science tends to "disappear" from the debates among lay people. Rather than debating their understanding of the science, people discussed the credibility of information and how to relate this to local knowledge and understanding. From research on risk communication it is well known that many people tend to focus on the credibility of spokespersons and the source of information rather than on the content of information (Frewer *et al.* 1996). To focus on credibility, competence, relevance and possible vested interests, or contextual factors implies assessing the sources of information. This contrasts with assessment using scientific criteria, which focuses on the content of the information they received.

In a study on students' investigation on issues about power lines (Kolstø 2000b) it became apparent that students at times focused on the content of the science-related information encountered, but without using the scientific criteria noted above. Some students explained that they would trust information when it looked "serious." Other students focused on whether the argument "looked logical" or not, whether it fitted in with what they already knew, and whether there was "proof" or not. In the study where student teachers assessed information related to a socioscientific issue, it was also found that many of them required the presence of references and scientific concepts to make an article appear trustworthy. These "reliability indicators"

(Kolstø 2000b) are used by most of us in our everyday lives when assessing credibility of information. One important weakness, however, is that such assessment criteria might be used to manipulate an audience. One can convince an audience by adding details and references and by using a clear and logical argument, even if the references are irrelevant. This is not to say that logic and consistency in arguments are unimportant, but they need to be used with a critical attitude and an awareness of their limitations.

Therefore, research indicates that students' knowledge of frontier science and of processes related to the development of scientific knowledge is not very good. There are also indications that students do not discuss the meaning of scientific concepts when debating socio-scientific issues. In assessing information it also appears that students do not use scientific criteria to any great extent. Also, when focusing on contextual factors they do not seem to know how to judge which researchers or experts are more competent than others. Therefore, the key question that arises from this research is: What type of knowledge could strengthen students' investigation and evaluation of the science dimension of environmental issues?

To understand and to be able to distinguish between both types of science knowledge ("core" and "frontier") is important in order to understand different environmental controversies. However, content knowledge *in* science is more heavily emphasized in schools than knowledge *about* science - how to use scientific concepts or how scientific knowledge is developed. Ryder (2001: 35) concludes, on the basis of inspection of a number of case studies of individuals interacting with science, that: "[o]verall, much of the science knowledge relevant to individuals in the case studies was *knowledge about science*, i.e. knowledge about the development and use of scientific knowledge rather than scientific knowledge itself" (emphasis in the original).

A FRAMEWORK FOR DEVELOPING KNOWLEDGE ABOUT SCIENCE

Knowledge about the nature of science appears to be a prerequisite for the ability to adequately investigate and evaluate the science in environmental issues. A possible framework for developing this knowledge can be based on an understanding that: (1) science is process based; (2) science is provisional; (3) science involves values; (4) science is an inseparable part of society; and (5) science is only one of several aspects of environmental issues.

Science is process based

An understanding of the difference between core and frontier science can serve as a tool to interpret the perceived disagreement between scientists as natural and necessary. It is important to make decisions related to environmental issues on the best knowledge available. Consequently, it is important that people, due to perceived expert disagreement, avoid losing trust in the science, and turn to relativism. By understanding the importance of peer-reviewed publication, criticism, and argument in science, students and citizens might judge core science as highly reliable knowledge when confronted with contested scientific claims. Knowledge about such processes in science might also diminish the tendency to interpret expert disagreement only in terms of perceived competence of the experts and vested interests.

It is to be hoped that people in decision-making situations will also want to understand the scientific basis for the expert disagreement and make their own assessment using both scientific and contextual criteria. An understanding of the role of consensus in science helps them to decide how much confidence to have in claims that have different levels of support by scientists. An example of this is the climate issue. Despite the fact that there are several dissenters, a great majority of climate researchers support the conclusions of the UN climate panel. Knowledge of the process involved in the collection of evidence, publication, and peer review is important to enable people, confronted with frontier science daily, to conclude that different kinds of science deserve different degrees of trust, and that consensual science is probably among our most reliable kinds of knowledge.

Science is provisional

Knowledge that science is provisional can serve as a tool to understand how perceptions of scientific uncertainty might influence different stakeholders' use of scientific evidence. The provisional nature of science follows from the description of science as a process, where core science is based on "an open system of understanding, that is mutable" (Pickett 1994: 182), and develops as a result of open argumentation among experts in each field. This system of understanding is based on scientific criteria evaluating evidence and models.

Also, students and citizens need to understand that we interpret nature through our senses, or instruments, and the logical concepts used to organize these data. Thus our scientific knowledge consists of models of reality and not reality itself. Science seeks to validate all scientific models with empirical data and, by definition, we cannot test the model against all data. This insight makes it possible to understand that one can always discover weaknesses in scientific models. An understanding of this aspect of science is an important tool when evaluating arguments or evidence presented by stakeholders involved in environmental controversies.

Another consequence is that the predictions of science sometimes fail. After the Chernobyl disaster, scientists informed authorities and sheep farmers in Cumbria that the levels of radioactive caesium in the sheep would soon decrease (Wynne 1996). As it turned out, the prediction failed and an indefinite ban was put on sheep sales, with serious consequences for the farmers. Only later was it discovered that one possible reason for the failed prediction was that the soil in the area was acid and peaty, while the prevailing scientific model was based on knowledge of alkaline clay soils.

Science involves values

Knowledge about some of the values central to science can also help to prevent misunderstanding and facilitate communication between scientists and the public. In science, the demand for empirical evidence is essential. Still, many citizens become frustrated when the evidence they bring forward in local issues is denoted as anecdotal and not rigorous enough. Irwin (1995) argued that citizens are typically concerned with local and specific situations. Science, on the contrary, aims at universally applicable theories and explanations. The historical success of science is probably to a great extent due to its emphasis on evidence and universality. However, anecdotal knowledge might point to the existence of a particular problem. Awareness of local and anecdotal knowledge as potentially relevant to scientific model-building can serve as a tool to increase one's self-confidence when trying to communicate local knowledge to experts. However, a knowledge of reasons why the scientific community values universal and statistical evidence can also serve as a tool for understanding why scientists sometimes dismiss local and anecdotal knowledge when this is appropriate.

Systematic skepticism is also valued in science. This value is probably a source of misinterpretations when politicians and the general public want clear-cut answers and scientists seem unwilling to give a set answer. Aikenhead (1985) describes how scientists in a Canadian court case were reluctant to draw conclusions as to whether a poor quality plant sample belonged to the specific species named in the law. The reason was that the scientists emphasized morphological characteristics that were impossible to identify due to the deteriorated condition of the sample. The scientists' reluctance to draw a conclusion was misconstrued by the judge as "hedging considerably." Aikenhead (1985) concludes that the judge "appeared to be uninformed with regard to the values that scientists cherish, particularly the value 'suspension of belief' - wait until sufficient evidence accumulates before making a decision. A complementary value, "suspension of disbelief" (Holton 1978) helps a scientist to keep working on an idea, in spite of preliminary contradictory evidence. Awareness of the value of evidence and skepticism can help citizens understand scientists' statements made in public forums.

Science is an inseparable part of society

Understanding that science is an inseparable part of society is of vital importance in citizens' investigation and evaluation of science-related information. Science is often presented as a neutral deliverer of objective knowledge for decision-making in social issues. However, science today is produced and used in a wide variety of industries and organizations. It is more than just basic research practiced at universities to fill in the gaps of a discipline's theoretical foundation. Modern science is also linked to social, political, industrial, or military needs (Aikenhead 1994).

Therefore, when students evaluate science-related information, they usually lack knowledge of the institutions where scientists work (Kolstø 2000b). Knowledge about the institutional aspects of science is probably a prerequisite for investigating and evaluating scientists' level of expertise or research interests. What does it take to be regarded as an expert? How can a scientist's stance within the scientific community be assessed? Under what constraints do scientists work in different industries or institutions? Knowledge of where and for whom scientists work might increase students' awareness that neutrality, even if highly valued, can seldom be taken for granted. Even if science values empirical evidence, it is still possible for researchers from different institutions to arrive at diverging conclusions.

Collingridge and Reeve (1986) argue that scientists involved in controversies tend to be more critical toward evidence supporting antagonists' arguments than toward evidence on which their own conclusions are based. For example, Geddis (1991) described the controversy between the United States and Canada on the source of acid rain. In this case, there was at first a lack of consensus on whether the evidence for the source of the acid rain was conclusive or not, due to the difference in demands for certainty by each party (who would benefit from the claim). The idea that demands for evidence, and level of certainty, may vary and that the demands can be influenced by institutional interests, can serve as a tool to interpret disagreements between scientists, governments, or citizens, without students' having to consider antagonists as dishonest, or "stupid" (Geddis 1991). This idea enables students, or citizens, to evaluate partisan and antagonist knowledge claims in more depth and context. It also makes it easier to respect the views of antagonists.

Science is only one of several aspects of environmental issues

Environmental education must recognize and reflect that science is but one of several kinds of knowledge that are relevant for thoughtful decision-making on environmental issues (Aikenhead 1985). Most issues involve economics, law, and ethics, to name but a few. In fact, decisions are sometimes made without scrutinizing the science dimension, but are often based on opinions regarding the potential consequences for the environment and for human beings, combined with personal values.

Being able to distinguish between statements expressing value judgments, such as "the amount of sulfur in petrol should be reduced," and statements meant to describe a fact or a situation, such as "it is now technologically possible to reduce the amount of sulfur in petrol," is important. This distinction can serve as a tool for examining and assessing inherent values (Kolstø 2001) or policy positions. Scientific expertise is restricted to describing what is, or predicting what might be. Suggestions for decisions and actions imply what ought to be done, i.e., what outcomes are valued. A relevant example is recommendations concerning risk assessments and acceptability of a risk. Citizens might decide to trust researchers' risk assessments, even if assessments involve choices where personal values or vested interests might have had an impact. However, deciding on the risk the public will accept is not a scientific question. Students and citizens need to be able to distinguish between scientific and value statements in arguments and policy positions.

DEVELOPING ATTITUDES AND SKILLS FOR THOUGHTFUL DECISION-MAKING

The teaching of thoughtful decision-making in environmental issues requires more than a general understanding of the nature of science and science–society interactions. In addition to the knowledge about science, thoughtful decision-making also requires development of attitudes and skills. Attitudes include appreciation of scientific information, evidence, and arguments, while skills enable investigation, evaluation, and argumentation. A challenge for educators is how to best facilitate students' development of such attitudes and skills.

Modern learning theory suggests that learning is closely linked to the contexts in which the understanding, attitudes, and skills were acquired (Lave and Wagner 1991). However, cognitive psychology has found little evidence of people's ability to apply knowledge gained in one context to problems encountered in another (Kirshner and Whitson 1977). For example, students may be able to learn and reproduce correct explanations of scientific concepts in school contexts, but hold and apply misconceptions in everyday situations outside school (Solomon 1983). Learning theory also claims that attitudes and skills are most efficiently acquired when learners take part in activities where these competencies can be applied and when the students can learn from more competent practitioners. This "apprenticeship" view of learning indicates that, if we want students to acquire the necessary skills to investigate, evaluate, and engage in debates, and to do this based on an understanding of the nature of science, we need to design learning situations where the students participate in these activities.

Consequently, valuing evidence, or developing a critical attitude, are values more efficiently developed through experience and practical application than through lectures. Students should be placed in situations where they have the necessary support to think or act critically. Situations need to be created where the students can experience the importance of basing an opinion or decision upon critically examined information, and experience the process of building arguments and decisions based on evidence. It is also important that they have experience in identifying strengths and weaknesses in their own thinking and in that of others, through argumentation.

These suggestions indicate that developing students' competence, attitudes, and skills in thoughtful decision-making requires carefully designed teaching approaches. In addition to providing experience, opportunities to apply knowledge about science need to be developed. However, to apply general knowledge in complex real world situations is not a trivial matter. Layton (1991) has argued that, in order to apply knowledge in practice, theoretical scientific knowledge has to be reworked. This involves a process of applying idealized models to complex and interdisciplinary real-world situations. Reworking theoretical knowledge is demanding and it is paramount that environmental education focuses on this process. It is not enough to focus on acquisition of knowledge, but application to a specific decision-making process also has to be part of the learning for thoughtful decision-making.

TEACHING APPROACHES

Therefore, in order to teach thoughtful decision-making, environmental issues have to be brought into the classroom (or students have to go to the environmental issues). One way of doing this is through a "problem-posing approach." Kortland (1996) used this approach when teaching decision-making about the waste issue. In Kortland's model the first phase is to induce a sense of purpose or motivation to investigate an issue. During this phase the teacher seeks to build on the students' assumed willingness to contribute to a better environment. The second phase is designed to help students become aware of their need to extend their knowledge base and state these needs in the form of questions for further investigation. In the third phase the students extend their knowledge by experimenting and/or gathering information. The fourth phase involves the application of the extended knowledge in the development of an argued point of view or decision. Finally, students are asked to reflect on their extended knowledge and decision-making process. It is through this last phase that students may become conscious of the process they have been through. After this process it should be easier for students to apply what they have learned about decision-making when dealing with other environmental issues. An advantage with Kortland's approach is its focus on developing extended knowledge about an issue. However, development of knowledge about the nature of science is missing. This probably makes the approach less relevant for environmental issues that involve controversial science, or conflicting information.

Ratcliffe (1996) proposed a second teaching approach. This approach consists of a framework for developing students' decision-making skills using six steps: identification of options, identification of criteria to use in the decision-making process, assessing information and information needs, surveying the identified options and identifying pros and cons, making a decision, and reviewing the process looking for possible improvements. A potential weakness of this framework is that it does not teach students how to critically assess and apply knowledge about the nature of science to environmental issues. Several other authors have argued that asking students to debate a socioscientific or environmental issue helps develop students' argumentation skills (see e.g., Simonneaux 2001). Fullick and Ratcliffe (1996) have also suggested several teaching approaches using debate on ethical aspects of socioscientific issues. Although interesting, these approaches also do not include a focus on the credibility of knowledge claims or a prerequisite knowledge of the nature of science.

A teaching approach proposed by this author (Kolstø 2000a) focuses on the need to assess information gathered in a critical manner. In this teaching approach students are divided into "expert groups" to gather information on different aspects of an issue. The issue and the topic areas of the expert groups are identified through a class discussion. The expert groups present their findings at a class conference. One special group, called the "laity group," has a specific duty to identify alternative actions for the issue. At the conference they listen to the presentations and ask scrutinizing questions. In the end the laity group has to come up with a consensual decision on a recommended action. The demand for consensus stimulates students to negotiate and look into each other's arguments. The teacher's role at the conference is to ask critical questions of the expert groups regarding the information they are putting forward in order to help the students to develop a critical approach. Building competence to examine environmental issues and assess information in this case is facilitated through the teacher's modeling to ask critical questions and a class-based effort to rework information in order to recommend a course of action for the chosen issue.

The approaches presented above do not include teaching focused on developing students' learning about the nature of scientific knowledge production, interpretation of models, or institutional aspects of science. One approach for teaching modeling and data interpretation has been presented by Schauble *et al.* (1995). The approach is based on Dewey's (1913) suggestion to start with a compelling design activity and gradually shift toward understanding underlying mechanisms. The students were asked to develop, test, and revise plans for a set of vessels that could carry construction materials up a river. The main goal was to encourage students to reflect on their own ideas and evaluate their emerging theories in the light of new evidence. Schauble *et al.* (1995) conclude that "experimentation in the science classroom can engage students in reasoning about relevant scientific concepts and ... about the nature of experimentation itself." The strength of this teaching approach is the focus on the process of theory-building through the formulation of tentative hypotheses. However, it does not include learning about the role of the scientific community in evaluating and criticizing knowledge claims from "the laboratory" and comparing them with the findings of other researchers.

Students need to learn about the provisional aspects of science, the role and importance of evidence and argument, and the difference between frontier science and core science. In order to develop understanding of these aspects of science, Kolstø and Mestad (2003) developed an approach where two science classes at two different schools carried out research on the same question. The question was: "What are the optimal living conditions for sowbugs?" The students worked in groups to select species (e.g., *Porcellio scaber*), specify a hypothesis, design a research method, put this method into practice, evaluate results, and write a report. The reports from all the groups were made available to everyone (through the use of a learning management system). Students were asked to assess the different reports and send comments to each other. Following this on-line discussion there was a class discussion on the credibility of the conclusions from different groups. Included in this discussion was a debate on the extent to which the whole process of experimenting and argumentation imitated science as a process. By organizing competing clusters of groups into "scientific institutions" or industrial companies, debates on the nature of institutional aspects in science were also initiated. The disadvantage of this approach is the complexity of learning objectives, making it demanding for the students to keep track of all the new ideas. There is clearly a need for more research on how to develop knowledge, attitudes, and skills for investigating and evaluating the science, and developing thoughtful decision-making regarding environmental issues.

CONCLUSION

This chapter has focused on a possible framework and teaching approaches to enhancing citizens' understanding of the science involved in environmental issues. Thoughtful decision-making requires an understanding of what is being debated, assessment of science-related knowledge claims, and debate with peers, in order to clarify personal views and to understand other points of view. Students and citizens tend to base personal views or decisions on the ethical aspects, possible risks, or source of information, rather than on the credibility or validity of the science or evidence. Therefore, environmental education needs to emphasize knowledge about science if we want students and citizens in general to investigate and evaluate controversial science-related claims made in environmental issues.

Such knowledge can provide "tools" which students can use according to their own agenda and personally held values, avoiding the idea that there is one "correct" way to examine and assess the science dimension of environmental issues. This might seem like a contradiction of this chapter's emphasis on the need for "adequate" assessment. However, to claim that there is more than one adequate way to perform evaluations is not the same as saying that all evaluation is equally valid. As in science, it is important that different perspectives are debated in public or with peers, making it possible to challenge views and interpretations regarded as inadequate or insufficient. Hopefully, such feedback will stimulate thorough consideration of one's own view and lead to rethinking or clarification of one's own argumentation.

Educators also need to create situations where the necessary skills of investigation and evaluation of information can be developed. For example, situations involving debate help students develop a positive attitude toward evidence and its use in argumentation. However, inclusion of scientific controversies and frontier science is not customary in compulsory school science education. Traditional science education focuses on core science, and this science may be delivered as indisputable facts. It should be noted that the teachers' ideology of knowledge, their view of authority and argumentation in teaching, and their willingness to make scientific debate manifest is crucial for the students' development of intellectual independence (Geddis 1991). By presenting a claim as an established fact, the possibility of critical scrutiny by students is pre-empted.

Students also tend to hold a naïve view of science wherein single measurements are believed to provide new scientific facts, and awareness of the role of publication and argumentation is absent (e.g., Lederman 1992). Textbooks presenting historical experiments and evidence as "proving" theoretical presumptions to be right or wrong probably contribute to this view. Thus there is a tension between the image of science inherent in most science courses today and the account of science given in this chapter. Consequently, in order to develop thoughtful decision-making on environmental issues, science education in schools needs to rethink its content and the image of science presented.

An emphasis on science as provisional and with varying degrees of reliability might offend some science educators whose view of the nature of science differs from the one which has been presented in this chapter. To include current controversies in science education and help students to arrive at reasoned views on such issues implies taking them to the door of political action. Making topical environmental controversies part of education, involving the analyses of the role of evidence, source of information, and level of consensus among scientists might provoke those who want education to be "neutral" and not to become politicized. Controversy, however, is an inherent part of environmental issues. Consequently, if we want environmental education to be relevant to the challenges confronting citizens today, and if we want citizens to carry out autonomous investigations and evaluations, the inclusion of a realistic image of science and topical controversies is paramount.

ACKNOWLEDGMENTS

I would like to thank Anne-Brit Fenner and Ole Einar Torkildsen from the Department of Applied Education at the University of Bergen, Bergen, Norway for reviewing earlier drafts of this chapter.

REFERENCES

Aikenhead, G. S. (1985). Collective decision making in the social context of science. *Science Education*, **69**(4), 453–475.

(1994). The social contract of science. In *Education. International Perspectives on Reform*, eds. J. Solomon and G. Aikenhead, pp. 11–20. New York, NY: Teachers College Press.

Barnes, B. (1987). Power listens to science. *Social Studies of Science*, **17**, 555–564.

Bingle, W. H. and Gaskell, P. J. (1994). Scientific literacy for decision-making and the social construction of scientific knowledge. *Science Education*, **72**(2), 185–201.

Brekke, P. and Anker-Nilssen, P. (1999). Med værforbehold om fremtidens klima. [With weather reservations for the future climate.] *Aftenposten* **12**. May.

Cole, S. (1992). *Making Science. Between Nature and Society*. Cambridge, MA: Harvard University Press.

Collingridge, D. and Reeve, C. (1986). *Science Speaks to Power: The Role of Experts in Policy Making*. London: Frances Pinter.

Dewey, J. (1913). *Interest and Effort in Education*. Boston, MD: Houghton Mifflin.

Fligge, M., Solanki, S. K., and Beer, J. (1999). Determination of solar cycle length variations using the continuous wavelet transform. *Astronomy and Astrophysics*, **346**(1), 313–321.

Frewer, L. J., Howard, C., Hedderley, D., and Shepherd, R. (1996). What determines trust in information about food-related risks? Underlying psychological constructs. *Risk Analysis*, **16**(4), 473–486.

Fullick, P. and Ratcliffe, M., eds. (1996). *Teaching Ethical Aspects of Science*. Southampton, UK: The Bassett Press.

Geddis, A. N. (1991). Improving the quality of science classroom discourse on controversial issues. *Science Education*, **75**(2), 169–183.

Holton, G. (1978). *The Scientific Imagination: Case Studies.* Cambridge: Cambridge University Press.

IPCC (Intergovernmental Panel on Climate Change). (2001). *IPCC Third Assessment Report: Climate Change 2001, Summary for Policymakers.* [Online] URL: http://www.ipcc.ch/pub/spm22-01.pdf (retrieved February 25, 2003).

Irwin, A. (1995). *Citizen Science: A Study of People, Expertise and Sustainable Development.* London: Routledge.

Irwin, A. and Wynne, B., eds. (1996). *Misunderstanding Science? The Public Reconstruction of Science and Technology.* Cambridge: Cambridge University Press.

Kaarbø, A. (1999). Drivhuseffekten øker. [The increasing greenhouse effect.] *Aftenposten* **13**. April.

Kirshner, D. and Whitson, J. A., eds. (1997). *Situated Cognition: Social, Semiotic, and Psychological Perspectives.* Mahwah, NJ: Lawrence Erlbaum.

Kolstø, S. D. (2000a). Consensus projects: teaching science for citizenship. *International Journal of Science Education,* **22**(6), 645–664.

(2000b). "To trust or not to trust ..." Pupils' ways of dealing with a socio-scientific issue. *International Journal of Science Education,* **23**(9), 877–901.

(2001). Scientific literacy for citizenship: tools for dealing with the science dimension of controversial socio-scientific issues. *Science Education,* **85**, 291–310.

(2003a). Et allmenndannende naturfag. Fagets betydning for demokratisk deltagelse. [Science education for citizenship. School science and democratic participation.] In *Naturfagdidaktikk. Perspektiver Forskning Utvikling,* ed. D. Jorde and B. Bungum. Oslo, Norway: Gyldendal Akademisk, pp. 59–85.

(2003b). Kritisk vurdering av naturvitenskapelige på stander i media. [Critical evaluation of scientific claims in media.] In *Proceedings from the Nordic Conference in Science Education,* Kristiansand, July 5–8, 2002, ed. E. K. Henriksen and M. Ødegaard.

Kolstø, S. D. and Mestad, I. (2003). Learning about the nature of scientific knowledge: the imitating science project. Paper presented at the ESERA Conference, The Netherlands, August 19–23.

Kortland, K. (1996). An STS case study about students' decision making on the waste issue. *Science Education,* **80**(6), 673–689.

Lave, J. and Wenger, E. (1991). *Situated Learning: Legitimate Peripheral Participation.* Cambridge: Cambridge University Press.

Layton, D. (1991). Science education and praxis: the relationship of school science to practical action. *Studies in Science Education,* **19**, 43–79.

Lederman, N. (1992). Students' and teachers' conceptions of the nature of science: a review of the research. *Journal of Research in Science Teaching,* **29**(4), 331–359.

Lewis, J., Leach, J., Wood-Robinson, C., and Driver, R. (1997). Genetic Engineering – The Limits, discussion by 15–16 year old students on the acceptable uses and limitations of genetic engineering. Paper presented at the ESERA Conference, Italy, Sept. 2–6.

Norris, S. P. (1995). Learning to live with scientific expertise: toward a theory of intellectual communalism for guiding science teaching. *Science Education,* **79**(2), 201–217.

Picket, S. T. A., Kolasa, J., and Jones, C. G. (1994). *Ecological Understanding: The Nature of Theory and The Theory of Nature.* San Diego, CA: Academic Press.

Ratcliffe, M. (1996). Adolescent decision-making, by individuals and groups, about science-related social issues. In *Research in Science Education in Europe,* ed. G. Welford, J. Osborne, and P. Scott. London: Falmer Press, pp. 126–140.

Ryder, J. (2001). Identifying science understanding for functional scientific literacy. *Studies in Science Education*, **35**, 1–44.

Schauble, L., Glaser, R., Duschl, R. A., Schulze, S., and John, J. (1995). Students' understanding of the objectives and procedures of experimentation in the science classroom. *Journal of the Learning Sciences*, **4**(2), 131–166.

Simonneaux, L. (2001). Role-play or debate to promote students' argumentation and justification on an issue in animal transgenesis. *International Journal of Science Education*, **23**(9), 903–928.

Solomon, J. (1983). Learning about energy: how pupils think in two domains. *European Journal of Science Education*, **5**(1), 49–59.

Wynne, B. (1996). Misunderstood misunderstandings: social identities and public uptake of science. In *Misunderstanding Science? The Public Reconstruction of Science and Technology*, ed. A. Irwin and B. Wynne. Cambridge: Cambridge University Press, pp. 19–46.

Part IV Integrating changing perspectives of ecology, education, and action

This last part provides examples of how educators can integrate the science of ecology and action within environmental education. Alan Berkowitz, Mary Ford, and Carol Brewer provide readers with a contemporary framework for integrating ecology and environmental education. Milton McClaren and Bill Hammond's chapter examines the nature of education and action in environmental education and provides a framework for learning about, through, and for action. These authors provide educators with two new frameworks for planning for learning about and through action in the context of interdisciplinary curricula. The section concludes with a chapter on environmental education programs based on citizen science and participatory action research presented by Marianne Krasny and Rick Bonney. Krasny and Bonney describe the goals and challenges of developing two programs that attempt to integrate science and education goals. The chapter concludes with insights on program development.

ALAN R. BERKOWITZ, MARY E. FORD, AND CAROL A. BREWER

11

A framework for integrating ecological literacy, civics literacy, and environmental citizenship in environmental education

INTRODUCTION

Environmental education practitioners span all of the natural and social sciences in terms of their training and passion. Practitioners have a range of science backgrounds, from very little science background to science degrees and some view science as a root cause of environmental problems. The matter is made more complicated by the fact that environmental education does not have a professional training dimension in the same way that physics or sociology do. Although such training gives disciplines focus and rigor, some say this leads to a rigidity that environmental education does not require. Thus, there is a diversity of perspectives on the role the science of ecology should play in environmental education.

There are two concerns about ecology in environmental education. First, the ecology can reflect outdated ecological science and epistemology. Second, at times, too much or too little attention is given to ecology, or there may be an overly rigid, linear, and hierarchical view of environmental education building on science (ecology) "facts" toward "environmentally responsible behaviors." There are several reasons why these problems arise and persist. Educators who emphasize "Environmental Literacy," or "Citizenship" as the primary mission of environmental education, especially those advocating the development of "pro-environment values," or "environmentally responsible behavior," can at times ignore, misinterpret, or take a strong oppositional stance towards the science component of environmental education. This problem is exacerbated by the marginalized nature of environmental education within formal education, the persistent challenges of scientific literacy, and the urgency of current environmental issues. This puts enormous pressure on educators to squeeze everything into environmental education, or to leave whole parts out. The lack of a socially sanctioned scope and sequence for environmental education makes educators wary that its goals can actually be accomplished.

This chapter will provide a perspective on the essential part that the science of ecology plays in environmental education. The first section will provide an overview of possible definitions, and relationships between *ecological literacy*, *civics literacy*, and *environmental citizenship*. The second section will provide a new framework for *ecological literacy* that can help guide future environmental education theory, research, and practice. The final section will identify possible pathways and challenges to integrating contemporary ecological science and the proposed ecological literacy framework in environmental education.

ECOLOGICAL LITERACY, CIVICS LITERACY, AND ENVIRONMENTAL CITIZENSHIP

Definitions and relationships

Environmental education involves two types of literacy - *ecological literacy* and *civics literacy*. *Ecological literacy* can be defined as the ability to use ecological understanding, thinking and habits of mind for living in, enjoying, and /or studying the environment, while *civics literacy* can be defined as the ability to use an understanding of social (political, economic, etc.) systems, skills and habits of mind for participating in and/or studying society. Being literate does not necessarily mean that one will act literate. The literate person may not read, the "financially literate" person may not balance his/her checkbook, and the ecologically literate person may not live more lightly on the earth. For this we need another term, *environmental citizenship* as the overall goal of environmental education. Thus, *environmental citizenship* can be defined as having the motivation, self-confidence, and awareness of one's values, and the practical wisdom and ability to put one's civics and ecological literacy into action. Environmental citizenship involves empowering people to have the knowledge, skills, and attitudes needed to identify their values and goals with respect to the environment and to act accordingly, based on the best knowledge of choices and consequences.

Explicit in our view of environmental citizenship are values, but we do not specify *what* these should be. We see it as inevitable that learners are exposed to values in all educational activities. The questions about values then become: (1) What and who determine which values students get exposed to – those of the teacher, the producer of the education material they're using, or socially agreed-upon values? (2) How will this exposure be carried out and placed in the fuller learning experience? (3) Will the values be made clear and explicit, and are the learners given the freedom, guidance, and opportunity to integrate the values (or not) into their own value system?

Perhaps the biggest challenge educators face is to move environmental education and its goal of environmental citizenship from a marginal to a more central position within education, social discourse, and public conscience. One manifestation of the marginality of environmental education is the propensity to include every facet of environmental education, no matter how limited the time. For example, a nature center running a single two-hour program might try to address all the dimensions of environmental education, thinking that this might be the only exposure to nature in a student's entire schooling. This puts tremendous pressure and constraints on our ability to elaborate a broad, comprehensive, and robust definition of the goals of environmental education. Another manifestation of the marginal position of environmental education is its focus on "problems." That is, much of what we dream of for humanity is not just the solution of problems, but also the generation of positive action and endeavors addressing human and non-human needs, values, and wants. Environmental citizenship should reflect this dream.

One of the most troubling consequences of environmental education's marginal position is that it fosters divisiveness and competition within the field. If we have only two hours for environmental education, what can we leave out? Is it the values piece, the science piece, or the action piece?' In explicit opposition to this tendency, we argue for an integrated environmental education whose goal is environmental citizenship.

An integrated framework for environmental citizenship

In order to address the central question of this chapter – what and how should ecology be taught within environmental education – we first must elaborate more completely a framework for *environmental*

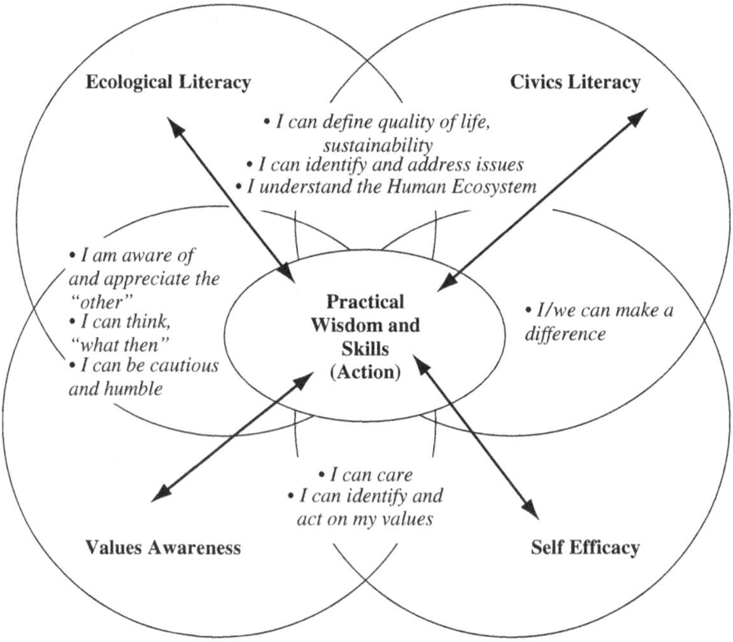

Figure 11.1. The five overlapping components of environmental citizenship. For each area of two-way overlap, illustrative statements of an environmental citizen are included. The prominent arrows emphasize the dynamic pathways by which literacy, awareness, and self-efficacy contribute to *and also benefit from* acquisition of practical wisdom and skills through action.

citizenship as a primary goal of environmental education. Figure 11.1 shows five overlapping components required for the development of environmental citizenship:

- *Ecological Literacy*: understanding the key ecological systems using sound ecological thinking, while also understanding the nature of ecological science and its interface with society.
- *Civics Literacy*: understanding the key social, economic, cultural, and political systems using the requisite critical thinking skills.
- *Values Awareness*: awareness of personal values with respect to the environment, and ability to connect these values with knowledge and practical wisdom in order to make decisions and act.
- *Self-efficacy*: having the capacity to learn and act with respect to personal values and interests in the environment.
- *Practical Wisdom*: possessing practical wisdom and skills for decision-making and acting with respect to the environment.

These components of environmental citizenship do *not* comprise a simple sequence, or hierarchy of steps, but are interrelated and highly overlapping. This framework makes explicit the idea that there are multiple "entry points" to teaching environmental citizenship. The arrows in Figure 11.1 emphasize the dynamic and two-way nature of interactions between action and the other components. This is in contrast to a more linear view of action as simply the culmination of knowledge and affective outcomes. The overlap zones between ecological literacy and each of the other components of environmental citizenship are rich areas for further development, now underexplored by environmental education. These overlap zones are personalized with statements of what environmental citizens might say in reflecting on their capabilities in each area.

While the bottom two components of the framework, Values Awareness and Self-efficacy, have a clear affective emphasis, there is a significant affective dimension to each of the other two components within this framework. For example, the disposition for self-reflection or metacognition and the ability to keep an open mind are essential parts of ecological and civics literacy. Also, cognitive skills identified in the North American Association for Environmental Education's (1999) definition of environmental literacy are subsumed within the two literacy components of this framework. The skills range from the ability to learn from experience and action, to thinking through one's feelings and values, to developing one's self-confidence to act and make a difference. Environmental literacy, as defined by NAAEE, might be viewed as the amalgam of the ecological and civics literacy components of this environmental citizenship framework.

Ecological literacy

A framework for this component will be described in detail in the next section.

Civics literacy

Civics literacy is the ability to understand social systems and to use this understanding to participate in society responsibly. While beyond the scope of this chapter to explore in much detail, we would like to raise the following questions about civics literacy. Are we being comprehensive enough in teaching civics literacy as defined above? Does environmental education address the importance of applying a whole-systems analysis to society?

The overlap between ecological literacy and civics literacy is a new dimension worthy of considerable attention and development. Perhaps this overlap zone replaces the third "knowledge" dimension in the NAAEE *Guidelines for Excellence* definition of environmental education, i.e., "Knowledge of Environmental Issues" (NAAEE 1999). Environmental issues and problems should be seen as essential topics for civics *and* ecological literacy, rather than as a separate dimension of environmental citizenship.

Values awareness

In order to foster environmental citizenship, people need the opportunity to develop and be aware of their own values having to do with other people and other organisms next door, up and down wind, up and down hill, up and down stream, and in the present and in the future. Environmental citizenship should build on the socially accepted value that the environment is a crucial consideration when making decisions and acting, rather than trying to agree on specific ways of valuing the environment or resolving the trade-offs individuals and societies face between competing values of the environment. This consideration is based on a recognition that humanity's health and happiness is inextricably linked, albeit in complex, contradictory, and confusing ways, to the status of the living and physical world around us. Environmental citizenship is the propensity to develop our values in this regard and act on them as we choose. It is also a process that evolves as our relationship with and understanding of the environment evolve, as our abilities as individuals and societies evolve, and as the world around us changes.

Self-efficacy

The development of understanding and skills, exploring and developing one's values, applying these to action, and using lessons from one's actions to further develop understanding and values, all require self-confidence. The connection between self-efficacy and the other components of environmental citizenship is a very intriguing and fruitful area of future research. How do ecological and civics literacy contribute to developing a sense of self-confidence in understanding the environment, and how do confidence, in turn, influence one's ability to learn and to make choices in the environment? How does self-efficacy grow from taking action and building practical wisdom and insight, and in what ways do low levels of self-efficacy or disempowerment limit action? Interestingly, these questions reemphasize the two-way interactions among components of the environmental citizenship framework.

Practical wisdom and skills

Environmental citizenship requires that one possess practical wisdom and skills for using knowledge, self-awareness, and self-confidence for action. It is through the application of practical wisdom and skills to real circumstances that people sharpen their understanding, develop self-efficacy, identify and clarify values, and reap the rewards and frustrations of citizenship. Taking action, and other expressions of environmentally responsible behavior, are listed as goals in many visions of environmental literacy. Our framework for environmental citizenship differs from these positions in two ways. First, the ability to take action is placed at the center of an interactive and iterative process among the components of environmental citizenship framework, rather than as an endpoint. This approach is similar to Shulman's (2002) taxonomy of learning, wherein action is considered "the pivot point, one might argue, around which most of education revolves." Second, we intentionally omit a values statement about the kinds of action environmental citizenship enables (e.g., "responsible, environment-friendly, sustainable"), to highlight our conviction that educational efforts to foster citizenship should be open-ended with respect to the specific values that are fostered.

The skills and practical wisdom needed to be an effective environmental citizen should build on knowledge from the ecological and civics literacy components. Practical skills are well described in existing standards (e.g., NAAEE 1999) and include skills such as problem-solving, issues clarification, communication, and persuasion. However, the specific arenas in which people choose to act, through community organizations, the political process, the legal system, policy and management, or individual behavior, leave open a broad range of practical skills needed, with each individual choosing his/her own path in this regard.

A NEW FRAMEWORK FOR ECOLOGICAL LITERACY

As part of this vision for environmental citizenship there is a need to develop a new conceptual framework for ecological literacy that will

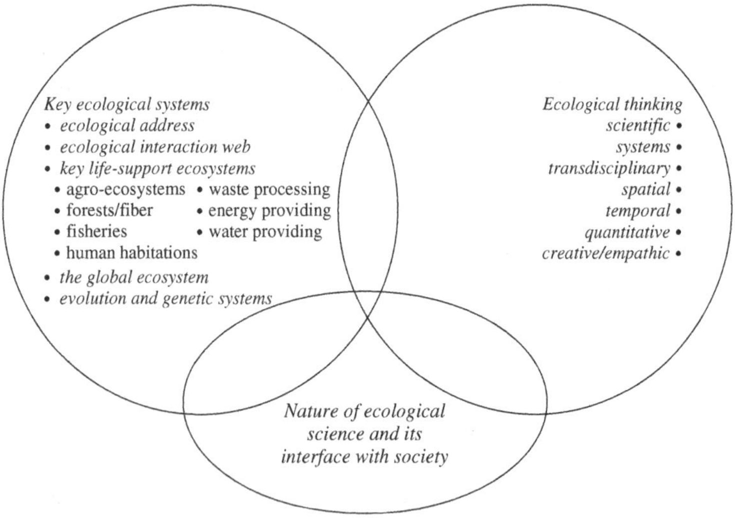

Figure 11.2. The three dimensions of ecological literacy.

provide clear guidance for environmental education theory, research, and practice. In our effort to craft a pedagogy-guiding view of ecological literacy, we struggled with the same challenges that previous efforts faced: balancing brevity versus comprehensiveness; assuring practicality while still trying to be provocative and inspirational; being synthetic and novel while sufficiently reflecting current vernacular. We are proposing a framework that we hope is novel and intriguing enough that it will spur debate, discussion, and reformulation.

In this effort we were significantly inspired by Paul Risser (1986), who, in his Past President's Address at the Annual Meeting of the Ecological Society of America, presented a definition of ecological literacy that took seriously the challenge of brevity. Risser's definition included just four things that everyone should understand: (1) multimedia transport of materials; (2) clarifying the "everything is connected to everything" concept; (3) ecology–culture interactions; and (4) familiar ecological field observations based on a specific, local "spot." We join Risser in accepting that we cannot teach everyone the whole discipline of ecology and the other environmental sciences, and ask, again, what is the shorter list of key things people should know and be able to do? The new framework we propose (Figure 11.2) has three components:

- Constructing an understanding of five key ecological systems.
- Building the disposition, skills, and capacity for ecological thinking.
- Developing an understanding of the nature of ecological science and its interface with society.

Five key ecological systems

What are the essential things people need to understand about ecological systems in order to be ecologically literate; i.e., what are the key ecological systems which merit understanding? This is where our thinking departs most significantly from the current frameworks used for ecology (summarized in Pickett *et al.* 1994). Rather than structuring our framework for ecological literacy on the traditional sub-field structure of the discipline of ecology – autecology, populations, communities, ecosystems – we are proposing that for the general learner, as opposed to the future ecologist or environmental scientist, there is a need for something different. Specifically, there is a need to understand, using the tools of *ecological thinking* (defined in the next section), the following five key ecological systems:

- One's ecological neighborhood or ecological address – one's home community and ecosystem.
- The ecological basis of human existence – the ecological connections that sustain and are impacted by us.
- The ecology of the systems that sustain us – e.g., the food-, fiber-, and other ecosystem service-providing systems we depend upon.
- The globe as an ecosystem and our impacts on it.
- Genetic/evolutionary systems.

There is significant and important overlap among all of these systems, reflecting the interrelated nature of the science of ecology. Another feature of the systems is the pervasive place of people, in contrast with many traditional views of ecology that exclude humans or treat them as external influences on the systems of interest. The science of ecology is expanding its scope of legitimate inquiry to include humans as parts of ecological systems, including managed systems, settlements, and engineered systems. Ecology is starting to define and understand new ideas, such as ecosystem services, and links with the social sciences and humanities. As such, environmental education also needs to address these five key systems more directly, critically, and thoroughly.

Ecological neighborhood or ecological address

What are the physical, biological, and social systems right in one's immediate environment, and how do they interact with and affect us? Ecological neighborhood includes developing a sense of place, an understanding of how one is connected to the local environment, and how it affects and is affected by the people living there. It requires familiarity with some of the more common and unique organisms, biological communities, soils, landforms, climate, human institutions, and cultures near one's home. A multidimensional community perspective of one's home is the ultimate goal here, with a basic appreciation of the interplay between environmental conditions and organisms and how this interplay helps explain the distribution, abundance, and diversity of organisms in the immediate environment.

The ecological basis of human existence

What are the essential movements of matter and energy that sustain humans and connect us with both local *and* distant ecosystems? In a sense, this is similar to one of Risser's (1986) components of ecological literacy - understanding multimedia transport - but with the specific application to people's central role in gathering and dispersing materials both near and far. This means being able to answer: Where do my food, water, energy, and oxygen come from, and where do my wastes go? While many of the answers to these questions are part of understanding one's ecological address, many of our resources and connections are quite distant both in space and in time, requiring a much more distributed understanding that complements an appreciation of our local ecology. It also means being able to answer: Why are there the numbers of people in my home, region, country, or the world, and what are the consequences of these numbers on the flow of energy and matter at each scale?

The ecology of the systems that sustain us

While the previous component emphasizes the ability to understand the flows of material and energy from and to people, this component focuses on understanding the ecosystems that sustain human existence. Specifically, people should understand the following systems: (1) agro-ecosystems; (2) fiber-producing ecosystems (e.g., forests); (3) fish-producing ecosystems; (4) waste-assimilating ecosystems (septic systems, sewage treatment/aquatic ecosystems, landfills); (5) clean water-providing systems (groundwater, surface water, etc.); (6) energy-yielding systems (e.g., biomass production systems, wind farms, hydroelectric plants, tidal flow harnessing); and (7) human settlements (cities, suburbs, etc.) (Berkowitz *et al.* 2003). The specific systems one needs to know about will vary from place to place; people should understand *at least* the systems that they, personally, rely on. The term ecosystem services, receiving increasing attention within science and education recently, provides another shorthand for this idea. In this framework, we emphasize the need to understand the source areas of these services *as ecosystems*, rather than focusing on the services themselves. For example, rather than simply appreciating the value that near-shore ecosystems provide to people through their use as fish farms, we suggest people need to understand these fish farms as ecosystems. This means applying spatial and temporal systems thinking (as defined in the next section on ecological thinking) to understand the key components of these farms and the important connections that these enterprises have with the systems that contain and are adjacent to them. This is one of the prominent places where this framework for ecological literacy connects directly to education for sustainability.

The globe as an ecosystem

People need also to understand the globe as an ecosystem, appreciating the ways that the laws of conservation of mass and energy function at the global scale and the implications for life on earth. Interestingly, an understanding of the globe's atmosphere and the fundamental roles of plants, heterotrophs, and people in shaping it, is a special case of a "key system that sustains us" as described in the preceding paragraph. Since humans are impacting the entire globe through atmospheric alteration, land surface conversion, long-distance movement of species, etc., people need to know the basic flows and their consequences. An appreciation of the magnitude of our impacts and the time lags involved in current and projected changes is essential.

Genetic/evolutionary systems

How do genes operate in ecological systems, how and why does the genetic make-up of populations change over time through evolution, and what is the significance of evolutionary processes in helping us understand the other facets of ecological systems identified in this framework? This part of ecological literacy lies at the interface between the theory of evolution and our understanding of the environment. People need to know what the basic sources of heritable variation in populations are and how natural selection works on this variation. They also need to understand the tension between the persistence of traits and what can be either the slow or rapid change of traits through time. Finally, they need to understand how humans, intentionally and inadvertently, use selection to affect both the generation of variation and the selection of traits (e.g., by introducing novel chemicals into the environment, moving species and genes around the globe, etc.). An understanding of processes such as disease resistance to antibiotics, pest resistance to chemicals, and human genetic responses to environmental changes must be included in the development of an ecologically literate citizen.

Disposition, skills, and capacity for ecological thinking

How should people think about ecological systems? We propose to answer this question by identifying and describing an "ecological thinking toolkit" comprising seven essential components:

- Scientific or evidence-based thinking
- Systems thinking
- Trans-disciplinary thinking
- Spatial thinking
- Temporal thinking
- Quantitative thinking (emphasis on probabilities and uncertainties)
- Creative and empathic thinking

These components are related to the "scientific ways of knowing," thinking skills, critical thinking, and cross-cutting skills identified in the US National Science Education Standards (National Research Council 1996), the AAAS Project 2061 Benchmarks (American Association for the Advancement of Science 1993) and the questioning and investigation skills in the NAAEE *Guidelines for Learning* (NAAEE 1999). These ways of thinking also build on and are complementary to other frameworks used to describe scientific ways of knowing, e.g., the different reasoning modes of Hogan and Weathers (2003); important criteria for understanding defined by Mintzes *et al.* (2001); and different types of causality elaborated by Grotzer and Perkins (2000). However, there are three key reasons for identifying ways of knowing or thinking that are *specific to ecology*. First, we assert that thinking skills or ways of knowing are best learned within a specific domain and their application to a domain is best accomplished by teaching them within that domain. Developing domain-specific knowledge and more generalized thinking skills are linked (Schauble 1996; Mintzes *et al.* 1998). Second, we think that ecological thinking can be defined in a more clear and specific way than if left as more generic or across the sciences ways of thinking. Finally, we think that the implications for pedagogy and environmental education practice are more apparent and less subject to misinterpretation.

While beyond the scope of this chapter, it would be interesting to consider what kinds of advances in defining environmental education might come from developing a similar list of critical thinking components for civics literacy. If *understanding* of social systems was the goal, rather than simply enabling individual and collective behavior, then the list of thinking skills might be broader and more comprehensive than what is found in current frameworks.

Scientific or evidence-based thinking

Scientific thinking might most simply be considered evidence-based thinking. The ecologically literate person:

- Understands the different types of evidence needed to answer ecological questions;
- Understands and evaluates the different sources of evidence addressing ecological questions;
- Can think through, investigate, and participate in the collection and application of evidence to address questions they have about the environment.

Note that the pedagogy implicit in this component requires learning by doing. However, we suggest engaging students in doing research *not* so that they will become lifelong researchers, but so that they will become better scientific thinkers (i.e., more adept at grappling with evidence in all of their dealings). Students should learn how to evaluate different kinds of experimental evidence by conducting their own ecological experiments and by reflecting on the utility and limitations of the results, in order to apply this form of evidence-based thinking in other aspects of their daily lives (Finn *et al.* 2002).

Table 11.1 lists some of the important types of evidence in ecology and their various sources that people should know about. We should also provide guidelines for integrating and evaluating scientific thinking and types of evidence in ecology. For example, stronger scientific arguments often involve the use of more than one type of evidence, such as combining experimental with comparative evidence or modeling with careful quantitative description. Indeed, it might be true that the strongest arguments integrate all four types of evidence. While current trends may place a premium on experimental over other forms of evidence in science in general, it is important to help people understand the limitations of this approach and other forms of evidence, specifically in ecology. Thus, the ecologically literate person knows the basic strengths and limitations of the different types of evidence, as well as the pros and cons of the different sources of evidence they might use or see used in public discourse.

Systems thinking

Systems thinking is a "contextual module" of thinking skills, knowledge, dispositions, and habits of mind (Bereiter 1990; Hogan and Weathers 2003) central to ecological thinking. Through systems thinking an ecologically literate person can:

- Define an object of study in the environment as a system with all the key components and their connections specified and bounded in time and space;
- Identify the two main types of (often overlapping) systems in ecology, those involving individuals, populations, genes, and evolution and those involving groups of species, communities, and ecosystems in functional ecological time;
- Place whole systems into their hierarchical context, using reductionism to look at mechanisms, comparisons across space and time at the same level to look at structure/function relationships, and holism to understand the role of the system in the systems that contain them;
- Understand the nature of causal factors, constraints, and feedbacks in ecological systems.

There is considerable interest in the teaching of systems thinking in ecology (Keiny *et al.* 2003; Smith 2003). Systems thinking provides a

Table 11.1. *Dimensions of scientific thinking in ecology: types and sources of evidence the ecologically literate person should be familiar with and adept at using*

Types of evidence			
Descriptive	**Comparative**	**Experimental**	**Modeling**
	• space-for-time substitution	• microcosms and mesocosms	
• monitoring • natural history observations • preliminary data • long-term studies	• gradients, real or conceptual • patches • natural experiments	• controlled studies • semi-controlled • time-for-space substitution • social action and management as experiments	• mathematical/deterministic • simulation/stochastic • physical models
Sources of evidence			
• direct investigation, data collection			
• primary literature, direct communication with research scientists, use of available data			
• secondary literature, intermediaries, hearsay			

framework for people to cope with the diverse scales and types of ecological systems and a vehicle for integrating understandings from the diverse disciplines that apply to the environment (e.g., physiology, chemistry, physics). Together with quantitative-probabilistic and transdisciplinary thinking, systems thinking helps integrate internal and external contingent factors, since ecological systems have multiple causality and complexity. Rigorous systems thinking provides a productive alternative to being too holistic or too reductionist, forcing us to be comprehensive in our understanding. For example, the recent emphasis within the field of ecology on including humans, conceptually, as part of ecosystems (e.g., McDonnell and Pickett 1993), is an example of using a rigorous specification of the earth's systems by recognizing that there is no place on earth free of human influence. In applying systems thinking in ecology, we must be careful not to romanticize systems or to misapply notions of self-assembly or self-regulation where there is no evidence for these features. Hogan and Weathers (2003) elaborated a more thorough treatment of systems thinking as a goal for ecology education.

Transdisciplinary thinking

A scientific understanding of the environment requires facility in linking perspectives from disparate disciplines. Thus, the ecologically literate person:

- Is able to apply understandings of the environment from the other natural sciences (physics, chemistry, geology, hydrology, meteorology, mathematics, and the social sciences) to ecological phenomena;
- Possesses a working framework for how to use the disciplines that surround and interlink with ecology for understanding;
- Has the dispositions and personal traits needed to reach out and to collaborate with people in other disciplines.

The ability to apply other disciplines to environmental questions requires a number of skills and attributes. There is, unfortunately, no substitute for understanding the specific ways of thinking from these other disciplines that are most essential for ecology. For example, from physics – the ecologically literate person must be comfortable with the basic laws of thermodynamics and states of matter; or from chemistry – some basic acid/base and Redox chemistry or the unique features of water and a few other key substances. As in ecological literacy, literacy in these other areas must hinge on a few key ideas and ways of thinking, and these contribute to trans-disciplinary thinking in ecology. Here is an obvious place where these more discipline-specific frameworks dovetail with the broader scientific ways of knowing frameworks that provide people with a knowledge structure for synthesis and integration (e.g., Grotzer and Bell Basca 2003). Finally, in order to have the confidence and motivation to think across disciplines, one needs a flexible mind, self-confidence, enthusiasm for going beyond one's own expertise or the topic at hand, and the inclination and ability to learn from and get along with others.

Temporal thinking

Ecological phenomena are on the one hand quite often immediate, with recent conditions often dominating the current state of a system. On the other hand, we know that longer-term legacies and lags can be significant. Thus, the ecologically literate person must be able to:

- Think about the two principal time scales at work in ecology, historical/ecological time (years to millennia) and evolutionary time (centuries to eons), taking a long and very long view of the past and into the future;
- Expect lags and legacies from the past and anticipate their effects on current systems and future trajectories;
- Engage in "selection thinking" (Orians 2000; Freeman and Herron 2004) in order to understand how organisms change through time via natural selection;
- Think about both "time's arrow" and "time's cycle" (Gould 1987), understanding the basic patterns of change, constancy, repetition, and unique events.

To be an adept temporal thinker takes imagination, especially when it comes to an appreciation of the role of disturbance in ecological time or natural selection in evolutionary time. Such understanding is made more challenging by the fact that neither set of processes are easily visualized in class or semester time (Grotzer and Bell Basca 2003; Dodick and Orion 2003). Temporal thinking as part of ecological literacy requires a basic familiarity with the time scales of ecological, geological, and evolutionary processes, and a sense of when ecological and evolutionary time scales and processes overlap. For example, when do organisms evolve significantly in ecological time, such that evolution is a current concern (e.g., diseases, weeds, etc.)? By understanding the nature of evolutionary legacies, or the "ghosts of selection past," and what determines how long-lived ghosts are, people can understand key ideas relating to trade-offs and constraints.

Spatial thinking

The ecologically literate person:

- Understands how location in space determines the quality of any place;
- Is accustomed to applying the notion that most ecological interactions are most intense among adjacent entities at many scales, and also knows when this may not be true;
- Is adept at seeing how the environment changes over space, for example as gradients or patches, and the ecological causes and consequences of these patterns;
- Knows that the surrounding environment or neighborhood can override internal processes in shaping ecological systems;
- Understands his/her role as an observer on a particular scale and can think beyond this scale to smaller and larger scales.

Ecological systems are, for the most part, physically or geographically bounded. Whether we are talking about an individual organism, a community, an ecosystem, or landscape, all exist in an explicit area in space. Thus, ecological literacy requires being able to understand how things operate in space – identifying boundaries, patterns, causes, and consequences of change, and how the location of any given spot on the globe affects what happens there. To complement the overall generalization that interactions and effects are location- and scale-dependent, ecological literacy demands that one also understands that not *all* interactions follow simple spatial rules. For example, some nearby players are more tightly linked than others, and some important connections span large and hard-to-predict distances. The habits of mind involved in thinking about space include a spatial imagination and the ability to apply basic principles from physiology and the physical sciences to ecological entities. It also involves having the disposition to look beyond obvious patterns for less obvious causes and consequences. Spatial thinking overlaps with most of the other forms of ecological thinking – systems thinking, temporal thinking, scientific thinking – and also with the systems of interest component, most obviously, as an essential tool for understanding our own ecological neighborhood and the ecosystems that sustain us. Exciting tools and approaches are available for teaching spatial thinking (e.g., Gauvain 1993; Gould 1994) and its application to landscapes (Gergel and Turner 2002).

Quantitative thinking about probabilities and uncertainty

Ecological phenomena are, for the most part, stochastic or probabilistic. Thus, the ecologically literate person:

- Appreciates the nature and basic sources of variability in ecological processes and controlling factors;
- Understands how to use the basic ideas of probability in coping with the stochastic and highly variable nature of ecological systems;
- Can think and act "with uncertainty in mind."

Basic quantitative skills are essential for all forms of scientific literacy and perhaps even more so for ecology in light of the complex and variable nature of the systems and processes of interest. The key is to give people useful tools for coping with this complexity, including models (Grosslight *et al.* 1991; Ewing *et al.* 2003), the tools they need to relate models to the real world (Hilborn and Mangel 1997; Best 2001), and basic skills in statistics and probabilities (Cothron *et al.* 1989; Maret and Zemba 1997; Gross 2000; Kugler *et al.* 2003). At the same time,we need to build people's appreciation for the limitations of different kinds of statistical arguments. Finally, people need to develop the dispositions and habits of mind that will allow them to make the best judgments they can, even in the face of uncertainty, balanced by the requisite humility and open-mindedness needed to learn from experiments and mistakes.

Creative and empathic thinking

The ecologically literate person:

- Thinks creatively about environmental questions;
- Uses their empathy with other organisms and their human insights about systems that are on the same scale as human experience to better understand the environment;
- Knows how to guard against the pitfalls of subjectivity and teleology in scientific thinking, even while using creative and empathic thinking;

- Enjoys the fundamentally generative, creative process of building meaning and understanding (i.e., sees knowledge building as a satisfying and creative act).

Science involves a particular form of creativity on the part of its practitioners, one that benefits from the individual's use of metaphor, analogy, intuition, and insight, but where the products of these creative acts are regulated through other forms of critical thought and social construction. There are several obvious sources of creative thought in ecology that are worthy goals for ecology instruction (Fisher 1997). One involves a type of transdisciplinary thought, where conceptual viewpoints and perspectives from other disciplines are tried out in ecology. The idea of "ecosystem engineers" (Jones *et al.* 1997) is an example of borrowing the engineering concept as a creative leap in thinking about organisms' roles as habitat modifiers in shaping ecosystems.

Empathy, spatial thinking, and temporal thinking all comprise what we might call one's "ecological imagination," and the ecologically literate person needs a well-developed imagination. The empathy we feel for other organisms is a key source of creative insight that can contribute to our ability to understand them. For many, this motivates us to understand ecological systems and the organisms they contain, building on our sense of concern and compassion for "the other." Of course, there are enormous pitfalls in the anthropomorphism involved in this sort of creative thought. However, rather than squelch the underlying empathy, we suggest that it is essential to recognize it and nurture it in the context of critical thought and safeguards of objectivity (Zohar and Ginossar 1998; Tamir and Zohar 2004). Indeed, forcing oneself to have empathy with organisms that are not familiar, or that do not elicit emotional empathetic feelings (i.e., imagining the environment through the "eyes" of others) can be a powerful form of inquiry; for example, taking the ant's or the microbe's view of the environment. By making empathy an explicit target for instruction, we are more likely to develop it appropriately as a tool with clear safeguards, since people will think and feel based on empathy whether we want them to or not (Kellert and Wilson 1993).

Integrating the seven components of ecological thinking

The different modes of thinking proposed here are the thinking skills and approaches needed to go beyond the basic conceptualization of ecological thinking that "everything is connected," as posed by Muir (1911) and Orr (1993). Systems and hierarchy thinking, quantitative and probability thinking, spatial, temporal, and trans-disciplinary thinking all come to bear on this issue, helping us figure out just which things are connected in important ways. Ecological thinking, as a whole, builds on the notions that people's actions shape ecosystems, both because we are components within most ecosystems and because we are important agents of their negative as well as their positive changes (Keiny 2002). It recognizes that people's values shape their interactions with the world, in terms of our mental conceptualizations of reality as well as our physical actions. It addresses our dual roles in ecosystems building on the assertions that reality is human perception, that knowledge is what the scientific community has accepted as the best interpretation of the world at any given moment, and that interactions played out in a social context lead us to have mutual respect for the "other" and to recognize their right to be different. In conclusion, we must strive to develop ecological thinking as an entire set of tools that individuals use when confronted with an environmental question or a need for an ecological perspective.

The nature of ecological science and its place in society

The third component of ecological literacy places environmental science into its social context. The ecologically literate person:

- Understands how ecological knowledge is constructed in society, how values can influence this process and how to safeguard against bias; this requires knowledge of the dynamic, social, open and conditional nature of ecological science (Pickett *et al.* 1994; Del Solar and Marone 2001);
- Understands how society, politics, and economics can influence the theories and practice of environmental science;
- Is able to apply or support the application of ecological understanding to social needs and problems;
- Has an ethical stance concerning responsibility to use their ecological literacy.

This component of ecological literacy provides a clear connection to the other parts of environmental citizenship: civics literacy, values awareness and self-efficacy. We suggest that ecologists and ecology educators would be well served to embrace this as *part* of their mandate so that they can take the lead in defining, shaping, and teaching ecology and its interface with society. Otherwise, defining this crucial interface and directing practice to foster understanding of ecology and its role might fall to those who have a less nuanced view of science and its interface with society. By helping people understand the nature and limitations of scientific explanations, our definition of ecological literacy contrasts with others that portray science as a source of static and immutable building blocks for environmentally responsible or sustainable behavior. Ecological understanding should not be construed as a product, nor should ecological science be represented as static and external to the learner, but rather as being dynamic and within the continued grasp of the learner.

A much more insidious problem that is involved here is the naïve notion that many have about the nature of science and ecology. These notions include the ideas that every question has opposing answers with scientists behind them, that science is completely subjective (e.g., radical relativism (Tauber 2001)), or that science and its intellectual tools are part of the problem we all face with the environment and therefore cannot be part of the solution. It is incumbent upon environmental educators to repudiate these naïve notions and replace them, as much as possible, with more sophisticated and productive views of science. Therefore, a working definition of ecological literacy must address this topic directly.

CHALLENGES AND PATHWAYS TO INTEGRATING ECOLOGICAL LITERACY AND ENVIRONMENTAL EDUCATION

By what pathways does ecological literacy, however we might define it, get into environmental education? By understanding the basic pathways, we hope to gain insight into why the gaps we have noted exist and what we might do about them. Three important pathways, or sets of flows, are involved: (1) people; (2) ideas; and (3) power and influence among the key components of the environmental science/education knowledge "system." How can we integrate this new framework of ecological literacy with environmental education by working through these pathways?

We hypothesize a number of reasons why the ecology in environmental education practice might not reflect the version of ecological literacy described above:

- Ecological literacy requires hard-to-develop thinking skills and understandings, made more difficult by the marginal place that environmental education holds in education systems;

- People who know ecology are not teaching it and those who are teaching environmental education did not learn modern ecology or do not have ready access to it; this is exacerbated in systems where people teach outside their specialty (by choice or edict);
- The ecology in texts, curriculum materials, and frameworks is almost necessarily outdated, especially texts at the secondary school level (i.e., texts based on texts);
- The environmental education knowledge system, like many human systems, is inherently conservative and resistant to change;
- Some practitioners in environmental education see ecology as providing building blocks for specific environmental actions, and thus are less interested in a dynamic view of ecology emphasizing thinking skills and scientific process.

Despite these challenges, there has been great progress in the evolution of environmental education and the integration of ecological science into its practice. Likewise, there are exciting and pressing needs for the future. This section will describe seven possible pathways to integrating ecological literacy and environmental education:

- Education research
- Pedagogy
- Instructional materials and curriculum scope and sequence
- K-12 teachers' and environmental educators' training and support
- Schools and other institutions
- Standards and assessment
- Academics' roles in integrating ecological literacy

Education research

We need new scholarship in education research. How do students and the general public learn ecology? How do people learn to think ecologically? What are the cognitive processes at work for the individual and how do social and other contexts shape learning in the diverse situations in which learning takes place? Research on learning, necessarily set in the context of innovative programs and curricula, will do much to advance the integration of ecological literacy into formal and non-formal education. A similar research agenda faces non-formal and media-based efforts aimed at general public education. These, however, might face the added challenge of the blurred distinction between education and influence, or even the wholesale abandonment of education completely by certain outreach campaigns. However, we think great progress could be made by coming to understand how members of the general public, or targeted groups (e.g., neighborhood leaders, environmental activists, etc.), learn to think ecologically and how they come to understand the key ecological systems they are in. The flow of ideas to people is complex, made so by the uneven proliferation of the worldwide web as information source and communication facilitator. The requisite research effort is different but complementary to the current work that measures factual environmental knowledge (National Environmental Education and Training Foundation 2001; Murphy 2002) in that it must be informed by and should likewise inform theory about cognitive and epistemological development. Consistent with our interactive and dynamic view of the environmental education knowledge system, we do not expect or suggest that education and cognitive research will serve as a simple "first step" in a linear chain of ecological literacy integration, but rather will interact with the other parts of the system to inform, inspire, and learn from practice.

Another key question we need to understand concerns what students or the general public already know about the key ecological systems we've identified as essential components of "Ecological Literacy." This is necessary to guide a constructivist approach to teaching and the development of age-appropriate instructional materials. However, it poses an interesting challenge that will come up again when we consider assessment and standardized testing - how do we measure understanding of very local and individualized things such as one's ecological address or the interaction web one is embedded in? While it might be helpful to know whether people can tell where most of the energy used in their country comes from, we argue that for the purpose of guiding the teaching of ecological literacy, it would be more useful to know whether a learner knows where their regional energy comes from. The obvious but hard-to-implement solution is to build partnerships between education researchers and ecologists who, in fact, know the local and particular, in order to craft education research projects that gauge genuine understanding of real places.

Pedagogy

What kinds of teaching strategies, or pedagogy, do we think will be needed to foster ecological literacy within environmental education? Any pedagogy aimed at fostering ways of thinking clearly must engage students in thinking. One of the important advantages of the ecological literacy framework we propose is that it makes explicit pedagogy of thinking and reflection (Feinsinger *et al.* 1997). Inquiry-based teaching must be augmented with other forms of active engagement in thinking - systems, spatial, temporal, and quantitative. Teaching scientific thinking must also include working with the many different sorts of data and evidence described in Table 11.1. Since students should be aware of the major goals of their learning and the ways these are to be developed and assessed, they should be aware of and regulate their own thinking (i.e., engage in metacognitive discourse). The adept ecological thinker must know when to engage specific tools or sets of tools. One of the consequences of this pedagogical stance is a variation on the theme of "less is more," wherein students will consider fewer concepts but be engaged in deeper thought about what they are learning.

Teaching for ecological literacy must have a strong constructivist base, making the need to know what students already know imperative (Novak and Gowin 1984; American Association for the Advancement of Science 1993; Mintzes *et al.* 1998; Novak 1998). From research on learning and cognition in ecology that identifies basic patterns of student thinking, and from embedded assessments during instruction that reveal specific mental constructs and conceptions held by students, teachers can devise strategies to help learners develop more robust and accurate understandings.

One of the major goals of instruction for ecological literacy is to apply the scientific thinking part of people's ecological thinking toolkit (Figure 11.2) to fostering their understanding of the five key ecological systems. Students should also learn how to collect, use, evaluate, and synthesize different kinds of evidence (Table 11.1) in order to understand their ecological address, the systems they depend upon, evolution, and genetic systems, etc. This should be done with data they collect themselves, with raw datasets from others, and through various types of synthesized evidence presented in the primary and soft literature. We are further intrigued by the idea of linking direct inquiry via data one collects oneself with comparative and long-term study, for example, comparing one's ecological home using climate data collected locally with that from schools in other biomes. The availability of data on the Internet, and electronic communication to exchange queries and findings makes this possible in new and exciting ways.

Teachers and instructional materials also should "model" or actively demonstrate ecological literacy. For students to learn how to think ecologically, they need to interact with scientists and educators who give life to the skills, dispositions, and understandings they are striving to attain. Thus, teachers must be ecological thinkers and understand the key ecological systems that comprise ecological literacy. Curricula should include a focus on local ecosystems, on real connections between learners and the environment, and on the actual service-providing ecosystems learners are linked to. For example, a school's resource use (water, energy, materials, air), its environmental footprint, and its structure's or landscape's derivation from the environment near and far should be made plain and accessible to students and their teachers.

We need to make the invisible visible and tangible for students if they are going to develop strong ecological thinking skills and come to understand the critical ecosystems they are part of. Fortunately, we know a lot about some of the major stumbling blocks (e.g., microbes, chemistry, slow or long-time processes, large-scale systems, and distant influences) in this regard and how they manifest and influence learning, and the effective means of addressing them. The availability of new tools provides optimism in this regard, e.g., inexpensive microscopes and those that can be used in museums and other settings, time lapse technology, computerized animations, GIS, air photo and satellite imagery for large-scale visualization of spatial patterns.

Finally, pedagogy must engage learners in *doing* ecology. Clearly, learners should have multiple opportunities to do ecological inquiry, producing and applying evidence of the types shown in Table 11.1 in their own investigations, while relating these to results from the scientific community at large. This must include repeated, in-depth and progressive exploration of the real organisms and environments in their immediate neighborhoods, along with comparisons across regions and worldwide where appropriate. But the action component of our pedagogy must go beyond scientific inquiry and include opportunities for people to engage in participatory and community-based approaches to research and learning about the environment. Such action is not just a critical source of ecological knowledge, but also plays crucial roles as a motivator for engagement and as a tool for assessment. A suite of challenges faces this form of pedagogy, such as maximizing ownership and commitment on the part of the learners while keeping the activity manageable by the educator and brokering individual versus group interests, especially when the group is diverse and might include students, teachers, community members, and scientists.

Instructional materials and curriculum scope and sequence

What kinds of materials, curricula, and scope and sequence do we need for ecological literacy? Clearly, teachers and other educators need curricula and support resources that focus on local ecosystems, on the actual connections that exist between learners and the environment, and on actual service-providing ecosystems that people are linked to. This calls for new kinds of education material and support resources, either replacing or complementing standardized textbooks and national curricula with localized materials, or localized versions of materials where possible. The pioneering work of Feinsinger and colleagues (Feinsinger and Minno 1995) in developing materials for northern Florida inspired a whole generation of new efforts to provide resources for teachers about the ecology and natural history of local and readily accessible organisms (e.g., Anon. 1993; Brewer 2002a, 2002b). New field guides are available (Stevenson *et al.* 2003), and new educational tools for helping people understand their own ecology are under development (e.g., Ecological Footprint, Wackernagle and Rees 1996). There has been a tremendous expansion of spatially referenced and accessed data about environmental conditions (e.g., Environmental Protection Agency 2004).

The five key ecological systems identified earlier in this chapter should provide students with very concrete and directly relevant topics of investigation and experiential learning. Sequences should start with local key service-providing ecosystems. The school, neighborhood, watershed, and region should, to a large extent, *be* the curriculum. Human habitations, be they cities, villages, or even individual homes, are one of the key ecosystems all students must learn about, and therefore the home-schoolyard-city ecosystem must be not just the setting but the subject of study.

We need guidance from education research about what would be a reasonable scope and sequence in order for students to develop a progression of concrete to abstract understandings of real ecosystems (neighborhood, life support, ecosystem service-providing, etc.) and ecological thinking skills. For example, the sequence could start with the ecological address and even with a focal organism, population, or community within the nearby ecosystem for learners to study. At first there might be less emphasis on distant parts of the environment or phenomena that cannot be observed directly. However, revealing critical but invisible agents and processes (microbes, chemical and physical transformations) should not be left too late in the scope and sequence. Indeed, we hypothesize that many of the most significant misconceptions and conceptual difficulties high school and college students (Green 1997; White 2000), and adults have in ecology (e.g., biotic nature of decomposition and nutrient cycling, accurate understanding of photosynthesis, primary production, and matter transformations) stem from a lack of attention to fundamental ecological processes early in education.

The Atlas developed by the American Association for the Advancement of Science (AAAS 2001) and the work of Barker and Slingsby (1998) are good examples of progress in this regard. The challenge is to develop learning sequences that flow from simple to more complicated without sowing the seeds for misconceptions. None of the dimensions of ecological thinking are completely beyond even the youngest learner; to the contrary, young people grapple with evidence, systems, space, and time throughout their lives. What we need to decide is when to make the details of thinking explicit to students, what kinds of tasks to engage them in, and what kinds of benchmarks to use to gauge their progress. We might, for instance, start with a single form of ecological thinking (e.g., spatial thinking) and apply it to simple systems and then build more sophisticated thinking skills and start to link these with the other forms of thinking. Again, the AAAS Atlas presents one example of this kind of developmental thinking applied to the idea of systems.

It is essential that we make the desired sequence of thinking skills *explicit* to learners and their parents, to teachers and administrators, and to political leaders who shape education policy. However, developing a scope and sequence for ecological literacy is particularly challenging since we have argued earlier in this chapter against a hierarchal and linear sequence. Thus, we should identify an appropriate number of "cycles" over the K-12 formal education through which environmental education can develop ecological literacy. The resultant scope and sequence for each component of ecological literacy might build on Schulman's (2002) taxonomy of learning: engagement and motivation, knowledge and understanding, performance and action, reflection and critique, judgment and design, and commitment and identity.

Ecological literacy education must also be integrated with other disciplines. For instance, students need to know certain things from basic biology, physics, and chemistry before they can apply these ideas to ecological systems. AAAS's Atlas project makes significant progress in this regard, including strand maps linking several sets of ecological ideas to those in other disciplines.

Teaching for environmental citizenship (Figure 11.1) requires a diversity of instructional models, with multiple entry points into a highly interactive process. Thus, ecology will not always come first, or in a set position within a rigid sequence. The goal is to achieve a robust, and diverse environmental education program at the school, school system, or state level, such that individual components or facets of environmental citizenship - such as a focus on ecological literacy, civics literacy, or values clarification - can be pursued in depth.

K-12 teachers' and environmental educators' training

The nature and quality of direct instruction of students and the general public by educators depends on: (1) Who is recruited into and retained in the education profession, and what training and knowledge do they receive before they start teaching? (2) How do substantive, epistemological, and pedagogical ideas flow to educators as they teach or communicate? (3) How do input and feedback from students, before, during, and after instruction, influence teaching? and (4) How do external factors and contexts influence educators' practice? Rigorous answers to these questions with respect to environmental education practice are beyond the scope of this chapter, but some issues can be identified to guide future research and practice.

The first important issue is the ability of educators to be strong ecological thinkers. However, are we training educators in this way in the K-12 system, or in colleges and universities? Even more vexing, are we recruiting people who are good ecological thinkers into the teaching professions or recruiting individuals who are less capable in this regard? For example, are K-5 teachers or environmental educators at nature centers stronger in civics and the affective domain of environmental citizenship and less strong scientific thinkers? If so, what are the implications in terms of their ability to teach ecological thinking by example?

The second important issue for teachers is the in-service training and support they receive. Where do teachers get their new ideas about the environment and about how to teach ecology? Some knowledge flows among teachers and from academic scientists and educators, while other sources include print and electronic resources that teachers acquire on their own. How can these pathways better support the integration of ecological thinking and literacy into environmental education teaching?

Third, the teaching practice is influenced by real or perceived input from the learners themselves. Teachers will benefit from the results of cognitive research about learners, especially if the information is made relevant and accessible. Teachers also need authentic assessment tools they can use to find out what their students know and what they learn about the new topics, along with the thinking skills we've identified as crucial for ecological literacy. The current movement towards developing instructional designs that embed authentic assessments strategically throughout the program will help and is especially challenging, but no less important, for the many environmental education programs that are short in duration (Moorcroft *et al.* 2000). Another key part revolves around teachers' ownership of and commitment to the goals of their instruction and to serving the full diversity of students they work with. Our vision of ecological literacy must evolve in at least two essential places in this regard. (1) There is a need for an overarching framework for environmental education, requiring collective construction and consensus so that there is some level of mutual agreement that what we're shooting for is both worthwhile and attainable. And (2) There is a need for a working definition of ecological literacy that individual teachers feel is relevant, important, and within their grasp and that of their students.

Finally, a complex diversity of other external factors influence teachers – the social dynamics of the school, the physical and community setting, school system, professional unions and associations, available instructional materials and resources, standards and the results of a full suite of assessments ranging from their own to high stakes standardized tests.

Many of the teacher-oriented strategies that are needed to integrate ecological literacy into environmental education are not unique to ecology or to environmental education. The strategies include: creating a culture of ongoing professional development; establishing communities of inquiry and education that include teachers, scientists, educators, and community members (Brewer 2002b); providing peer and scientist support; providing resources and the opportunity to use them; working to increase the number of teachers actually teaching in their field; and developing standards for formal or informal educator proficiency and training (e.g., National Research Council 1996; NAAEE 1999).

Schools and other institutions

Teaching for ecological literacy is shaped by the institutional context within which teaching and learning take place. For schools, nature centers, museums, and other non-school educational settings, this includes: the social and professional community of the institution and the systems it is contained within (e.g., school district, state, and national agencies; or regional systems of nature centers, or institutional context of museums or field stations, etc.); political and public power and control; and resources such as instructional resources, human resources in the surrounding community, and the bio-physical resources of the institution and its surroundings.

What can institutions do to foster ecological literacy with these factors in mind? Education institutions such as schools, school systems, and non-formal centers should embrace environmental citizenship as an overarching goal for education, on a par with other forms of citizenship. They should also define for themselves the components of environmental citizenship, including the critical literacies they seek to foster and an overall strategy for attaining it and assessing their success. As part of this commitment, institutions should create a professional culture that supports the kinds of teaching, professional development, and community engagement necessary to foster ecological literacy, civics literacy, and environmental citizenship. Since much of what students should learn about is the local environment, this will require building close connections with important actors in the local environmental scene, such as government agencies, non-governmental agencies, neighborhood groups, and academic scientists. School–community partnership initiatives are outstanding examples of creating a positive context for fostering ecological literacy, with service learning and other connections between knowledge and action being prominent examples having wide application in recent years.

As suggested in the pedagogy section above, educational institutions should develop ecological literacy by providing students, teachers, and others with access to information about how the institutions themselves function as ecological systems. Ideally, they would also serve as laboratories for students and teachers to experiment with applying their knowledge to making changes and then learning from these manipulations (Kobet 2003). Imagine a school or nature center that routinely measured its energy and water use, wastewater discharge, rain- and snow-water fluxes, or one that kept careful records of its landscaping activities, mowing and clipping removal, fertilizer inputs, or exotic plant introductions. These data could be made available to the community and embedded directly into the curriculum to help students understand the real systems of which they are part. Schoolyards, neighborhoods, and other local environmental resources should be utilized to their fullest extent, because this is the learner's ecological address and contains the nearby parts of the ecological interaction web of all learners.

Institutions play a vital role in providing access for teachers and students to the wealth of resources available through libraries, on the Internet, and in their greater community. Scientists at government agencies, colleges, and universities, and non-governmental organizations all have invaluable expertise and perspectives to contribute. It is at the level of the school or the school system that we really can achieve the kind of diversity of professionals and resources that is needed to carry out the diversity of teaching approaches needed to foster environmental citizenship. Thus, ecologically literate teachers can focus on their specialty, collaborating with and complementing work by others proficient in civics literacy, values clarification, building self-confidence or crafting educational and rewarding action projects. This is not to say that each specialist operates in isolation or without some comfort and facility in the full suite of environmental teaching approaches, but specialization should occur without others feeling their specialty is left out if the institution as a whole has an effective and diverse community of professionals.

Standards and accountability

Standards lie at the interface between education research, institutions, curriculum, instruction, and assessment. Ideally, standards should reflect the most current vision for what students should know and be able to do as informed by the interaction of these components. Standards should be useful for instruction, grounded in education theory and research, and linked to productive assessment tools for educators. Finally, standards are molded and prioritized by education institutions with important inputs from the public and political leaders. Input by professional associations of scientists often does not take place in a concerted fashion. The NAAEE, however, did seek input from professional science associations and has developed a very progressive set of standards for environmental education (NAAEE 1999) that are quite consistent with our vision for ecological literacy. However, what impact and role such non-mandatory national standards can and will play remains to be seen.

The next step in implementing our vision of ecological literacy will be to identify standards for participants in formal and non-formal education programs. This includes the challenge to develop assessments for gauging ecological literacy. Many of the intended learning outcomes are novel and complex, and others hinge around local and individual-based understandings that are difficult to look at through standardized tests. Likewise, it is difficult to assess the development and ability to use thinking skills, especially in the context of standardized tests (Blank and Brewer 2003), placing our framework for ecological literacy at or beyond the cutting edge of the assessment field. Many national and state standards include an emphasis on thinking skills, general or crosscutting dispositions or habits of mind (e.g., AAAS 1993; NRC 1996), and this framework can inform integration and synthesis of these skills. However, the success of translating these parts of the standards into state and local mandates, and into practice, with reliable assessment to back them up, has been much more challenging than the adoption of the more straightforward content and inquiry standards. However, it appears that the field of education is well aware of this challenge (Bybee 2003), and our framework for ecological literacy can be part of the cutting edge work in the future. The environmental education field needs to tackle this challenge head-on, developing ecological literacy metrics, and working to have them used by teachers, schools, and systems to provide feedback on how the framework helps guide practice.

There are pros and cons in the struggle to infuse ecological literacy more thoroughly and explicitly in national and state standards. While being omitted helps to perpetuate the marginal position of environmental education in the curriculum, being included opens it to be misconstrued and constrained. Standards that embody pedagogy, like our own definition of "Ecological Literacy," could be progressive if adopted. The AAAS 2061, NRC National Science Education Standards, and NAAEE *Guidelines for Learning* documents all emphasize a much broader range of standards than simply focusing on learning and learners. Those interested in fostering ecological literacy would be well advised to learn from the successes and the limitations of these efforts.

Another dimension of the standards and accountability issue revolves around equity. As long as environmental education and ecological literacy are *not* hard-wired into agreed-upon standards and programs, there will be inequalities in the delivery of environmental education programs and an asymmetry in the distribution of environmental citizenship among our population (Bryant and Callewaert

2003; Shu 2003). There is clear evidence that this is the case, with urban, poor, and minority populations often, though not always, evincing lower levels of environmental knowledge or awareness (Lee and Luykx 2003). This might be the most important reason to wage the battle for inclusion in high stakes standards and tests, linking to current efforts to avoid leaving any children behind. If parts of our populace continue to receive less instruction in ecological thinking, there will be dire consequences to society as a whole. Thus, resources and support for environmental education by institutions is a key dimension of environmental equity.

Academics' roles in integrating ecological literacy and environmental education

Academic scientists and educators, or, more specifically, ecologists and other natural and social scientists who study the environment, and education researchers who study teaching and learning, have multiple roles to play in integrating a modern version of ecological literacy into environmental education theory, research, and practice (Brewer 2001). These roles emerge for them as individuals, groups within their academic institutions, and in their disciplinary associations. The goals of their involvement should be to assure that environmental education has the tools it needs in terms of instructional materials, models, and other resources; is up to date and responsive to changes in scientific understanding of the environment and of how people learn; is responsive to societal changes; is balanced in terms of its emphasis, as a whole, on each critical facet of environmental citizenship; provides feedback and input into environmental research, decision-making, and management; and is given adequate attention and resources by society at large.

Who has power and influence to shape the way ecology and ecological literacy are integrated into environmental education? As a diffuse field, without an established place in either the overall curriculum or a unified professional identity, this is not easy to answer for environmental education. We suggest that ecologists and other academics must play a central role in this regard. Here, then, is a set of recommended actions for ecological scientists (again, broadly construed to include all natural and social scientists working in the environmental arena), including professional associations, to help address the need for a new framework of environmental literacy in environmental education. Ecologists and other ecological scientists are the best ones both to advocate for ecology in environmental education and to provide the frameworks and approaches needed to do so.

They can do this in many ways: through local and regional associations; by collaborating on projects to develop education standards, assessment, curricula, and training; and through their professional societies. Ecologists need to work with academic educators to shape the vision of ecological literacy in a pragmatic way and to define and advance a research agenda. As active educators in their own right (with undergraduate and graduate students, and in their work with teachers, students, managers, policy-makers, and the general public), environmental scientists can bring insights into developing innovative pedagogy, especially in light of the essential links between the key ideas in the field, ways of thinking, pedagogical content knowledge, and teaching. Thus, the content and pedagogy built into our working framework of ecological literacy and corresponding standards and assessments must be informed by scholarly natural science and cognition.

Ecologists and other scientists need to provide strong, positive models of excellent education practice to the fullest extent that they can. Many of the recommendations made in this chapter can easily be implemented in undergraduate teaching, where professors can focus on developing ecological literacy, addressing frameworks that emphasize key understandings and thinking skills, and using innovative and effective techniques that suit the full range of learners. A good number of their students will go on to be educators, and their future practice is shaped by the way they learn. Intensive research experience and internship programs provide, among other things, an opportunity for students and mentors to "think together." To the extent that academics can live and breathe ecological thinking, making their ways of looking at the world plain to their students and collaborators, they can help support others in developing these same understandings, knowledge, skills, and dispositions. This same notion of "modeling" ecological thinking should extend to the way that scientists structure the resources they produce for education, i.e., the texts (McComas 2003), websites, and curriculum supplements produced for formal and nonformal education use.

Ecologists are also in the best position to provide the kinds of local natural history and ecology knowledge needed to address ecological literacy as we envision it. The long legacy of nature study started over 100 years ago builds on a vital role for naturalists in the education process, and our modern vision of ecology education does not transcend the need for location-specific knowledge.

Finally, academics can provide a model for approaching ecological literacy through their very institutions. Our departments need to embrace trans-disciplinary thought and training, must support collaboration, and must honor a diversity of pathways to scholarly contribution if we are to have the kinds of programs in research and education that ecological literacy and environmental citizenship demand. Likewise, our professional associations need to transcend traditional boundaries and monolithic definitions of excellence to celebrate contributions from educators, social scientists, policy-makers, and researchers. Indeed, many of the professional associations, both research-oriented ones like the Ecological Society of America and educational ones like the National Science Teachers Association, are moving in this direction, giving us much optimism for the decade to come.

CONCLUSION

As humans' capacity to alter the environment reaches unprecedented levels, the urgency of fostering environmental citizenship in all people has never been greater and, perhaps, more difficult. Through specialization and urbanization, people find themselves further from the sources of their resources and less connected to the biological and non-human physical world. Though simple solutions might exist for some of the challenges we face, making simplistic behavior-oriented campaigns an option in these cases, for the most part the challenges are complex and difficult. Indeed, their resolution requires engagement and consideration by the full diversity of people involved, making environmental citizenship essential. In this chapter we have argued for a vision of environmental citizenship that is dynamic, involving an unfolding interplay between thinking, feeling, and acting through the course of one's life. Knowledge and understanding both feed and spring from one's actions. Values and a sense of self-efficacy contribute to and grow out of learning and doing. Perhaps most importantly, by making environmental citizenship an explicit and recognized goal of education, we raise each person's awareness of what it means to be a good environmental citizen. They, too, can then participate in the ongoing process of "inventing" the goals and vision of the environmental education enterprise.

One of the key components of environmental citizenship is ecological literacy. We have presented a framework for ecological literacy that is open and inclusive, and that involves the overlap and interaction between subjects of study (five key ecological systems) and ways of knowing (seven dimensions of ecological thinking), linked by innumerable areas of overlap, and by the additional emphasis on understanding how ecological science and society interact. Embodied in this view is a clear set of implications for practice: the need to make thinking skills and practical understandings of real ecological systems explicit to learners, teachers, and education systems; the need to engage teachers and learners in ecological thinking that builds their capacity and confidence over time; the need to provide local resources to support the teaching and assessment of learning about specific, local ecosystems and organisms; the need to carry out research in how people develop ecological thinking skills; the opportunity to capitalize on service learning, participatory research, and other action-oriented teaching approaches as sources of ecological understanding; and the challenge for ecologists to take an active role in assuring that their science is being taught - enough and well - in environmental education.

REFERENCES

American Association for the Advancement of Science. (1993). *Project 2061. Benchmarks for Science Literacy*. Washington, DC: American Association for the Advancement of Science.

(2001). *Atlas for Science Literacy*. New York, NY: Oxford University Press.

Anon. (1993). *Model Inquiries into Nature in The Schoolyard: An Inquiry Field Guide to the Natural History of Southwestern Virginia Schoolyards*. Blacksburg, VA: Virginia Tech Museum of Natural History.

Barker, S., and Slingsby, D. (1998). From nature table to niche: curriculum progression in ecological concepts. *International Journal of Science Education*, **20**, 479–486.

Bereiter, C. (1990). Aspects of an educational learning theory. *Review of Educational Research*, **60**, 603–624.

Berkowitz, A. R., Nilon, C. H., and Hollweg, K. S. (2003). *Understanding Urban Ecosystems: A New Frontier for Science and Education*. New York, NY: Springer-Verlag.

Best J. (2001). *Damned Lies and Statistics: Untangling Numbers from the Media, Politicians, and Activists*. Berkeley, CA: University of California.

Blank, L. and Brewer, C. A. (2003). Ecology education when no child is left behind. *Frontiers in Ecology and the Environment*, **7**, 383–384.

Brewer, C. A. (2001). Cultivating conservation literacy: "trickle down" education is not enough. *Conservation Biology*, **15**, 1203–1205.

(2002a). Conservation education partnerships in schoolyard laboratories: a call back to action. *Conservation Biology*, **16**, 577–579.

(2002b). Outreach and partnership programs for conservation education where endangered species conservation and research occur. *Conservation Biology*, **16**, 4–6.

Bryant, B. and Callewaert, J. (2003). Why is understanding urban ecosystems important to people concerned about environmental justice?

In *Understanding Urban Ecosystems: A New Frontier for Science and Education*, ed. A. R. Berkowitz, C. H. Nilon, and K. S. Hollweg New York, NY: Springer-Verlag, pp. 46–57.

Bybee, R. W. (2003). Ecology education when no child is left behind: forum response. *Frontiers in Ecology and the Environment*, **1**, 389–390.

Cothron, J. H., Giese, R. N., and Rezba, R. J. (1989). Simple principles of data analysis. *American Biology Teacher*, **51**, 426–428.

Del Solar, R. G. and Marone, L. (2001). The "freezing" of science: consequences of the dogmatic teaching of ecology. *BioScience*, **51**, 683–686.

Dodick, J. and Orion, N. (2003). Cognitive factors affecting student understanding of geologic time. *Journal of Research in Science Teaching*, **40**, 415–442.

Environmental Protection Agency. (2004). *Where you Live: Search your Community*. Washington, DC: US EPA. [Online] URL: http:// www.epa.gov/epahome/commsearch.htm.

Ewing, H., Hogan, K., Keesing, F., Bugmann, H., Berkowitz, A. R., Gross, L., Orvis, J., and Wright, J. (2003). The role of modeling in undergraduate education. In *The Role of Models in Ecosystem Science*, ed. C. D. Canham, J. J. Cole, and W. K. Laurenroth. Princeton, NJ: Princeton University Press, pp. 413–427.

Feinsinger P. and Minno, M. (1995). *Handbook to Schoolyard Plants and Animals of North Central Florida*. Gainesville, FL: The Nongame Wildlife Program, Florida Game and Fresh Water Fish Commission.

Feinsinger, P., Grajal, A., and Berkowitz, A. R. (1997). Some themes appropriate for schoolyard ecology and other hands-on ecology education. *Bulletin of the Ecological Society of America*, **78**, 144–146.

Finn, H., Maxwell, M., and Calver, M. (2002). Why does experimentation matter in teaching ecology? *Journal of Biological Education*, **36**, 158–163.

Fisher, S. G. (1997). Creativity, idea generation, and the functional morphology of streams. *Journal of the North American Benthological Society*, **16**, 305–318.

Freeman, S. and Herron, J. (2004). *Evolutionary Analysis*, 3rd edn. Upper Saddle River, NJ: Pearson / Prentice Hall.

Gauvain, M. (1993). The development of spatial thinking in everyday activity. *Developmental Review*, **13**, 92–121.

Gergel, S. E. and Turner, M. G. (2002). *Learning Landscape Ecology: A Practical Guide to Concepts and Techniques*. New York, NY: Springer-Verlag.

Gould, P. (1994). Perspectives and sensitivities: teaching as the creation of conditions of possibility for geographic thinking. *Journal of Geography in Higher Education*, **18**, 277–289.

Gould S. J. (1987). Time's arrow, time's cycle: myth and metaphor in the discovery of geological time. *Jerusalem-Harvard Lectures*. Cambridge, MA: Harvard University Press.

Green, D. W. (1997). Explaining and envisaging an ecological phenomenon. *British Journal of Psychology*, **88**, 199–217.

Gross, L. J. (2000). Education for a biocomplex future. *Science*, **288**, 807.

Grosslight, L., Unger, C., Jay, E., and Smith, C. L. (1991). Understanding models and their use in science: conceptions of middle and high school students and experts. *Journal of Research in Science Teaching*, **2**, 799–822.

Grotzer, T. A. and Bell Basca, B. (2003). How does grasping the underlying causal structures of ecosystems impact students' understanding? *Journal of Biological Education*, **38**, 16–29.

Grotzer, T. A. and Perkins, D. H. (2000). A taxonomy of causal models: the conceptual leaps between models and students' reflections on them. A paper presented at the Annual Meeting of the American Educational Research Association. New Orleans, LA.

Hilborn R. and Mangel, M. (1997). *The Ecological Detective: Confronting Models with Data.* New York, NY: Cambridge University Press.

Hogan, K. and Weathers, K. W. (2003). Psychological and ecological perspectives on the development of systems thinking. In *Understanding Urban Ecosystems: A New Frontier for Science and Education*, ed. A. R. Berkowitz, C. H. Nilon, and K. S. Hollweg. New York, NY: Springer-Verlag, pp. 233–260.

Jones, C. G., Lawton, J. H., and Shachak, M. (1997). Ecosystem engineering by organisms: why semantics matters. *Trends in Ecology and Evolution*, **12**, 275.

Keiny S. (2002). *Ecological Thinking. A New Approach to Educational Change.* Lanham, MD: University Press of America.

Keiny, S., Shachak, M., and Avriel-Avni, N. (2003). Teaching systems thinking: a model from Israel. In *Understanding Urban Ecosystems: A New Frontier for Science and Education*, ed. A. R. Berkowitz, C. H. Nilon, and K. S. Hollweg. New York, NY: Springer-Verlag, pp. 315–327.

Kellert, S. R. and Wilson, E. O. (1993). *The Biophilia Hypothesis.* Washington, DC: Island Press.

Kobet R. J. (2003) *Empowering Learning through Natural, Human and Building Ecologies.* Minneapolis, MN: The International Forum for Innovative Schools. Design Share.

Kugler, C., Hagen, J., and Singer, F. (2003). Teaching statistical thinking: providing a fundamental way of understanding the world. *Journal of College Science Training*, **22**, 434–439.

Lee, O. and Luykx, A. (2003). Ecology Education when no child is left behind: forum response. *Frontiers in Ecology and the Environment*, **1**, 384–385.

Maret, T. and Zemba, R. E. (1997). Statistics and hypothesis testing in biology: teaching students the relationship between statistical tests and scientific hypotheses. *Journal of College Science Teaching*, **26**, 283–285.

McComas, W. F. (2003). The nature of the ideal environmental science curriculum. Advocates, textbooks, and conclusions (Part II of II). *American Biology Teacher*, **65**, 171–178.

McDonnell, M. J. and S. T. A. Pickett. (1993). The application of the ecological gradient paradigm to the study of urban effects, in *Humans as Components of Ecosystems: the Ecology of Subtle Human Effects and Populated Areas*, ed. M. J. McDonnell and S. T. A. Pickett. New York, NY: Springer-Verlag, pp. 175–189.

Mintzes J. J., Wandersee, J. H., and Novak, J. D. (1998). *Teaching Science for Understanding: A Human Constructivist View.* San Diego, CA: Academic Press.

(2001). Assessing understanding in biology. *Journal of Biological Education*, **35**, 118–124.

Moorcroft, T. A., Desmarais, K., Hogan, K., and Berkowitz, A. R. (2000). Authentic assessment in the informal setting: how it can work for you. *Journal of Environmental Education*, **31**, 20–24.

Muir, J. (1911). *My First Summer in the Sierra.* Boston, MA: Houghton Mifflin.

Murphy, T. P. (2002). *The Minnesota Report Card on Environmental Literacy: A Benchmark Survey of Adult Education Knowledge, Attitudes and Behavior.* Hamline University and Minnesota Office of Environmental Assistance.

National Environmental Education and Training Foundation. (2001). *Roper Report 2000: Lessons from the Environment.* Washington, DC: National Environmental Education and Training Foundation. [Online] URL: http://www.neetf.org/roper/roper.shtm.

National Research Council. (1996). *National Science Education Standards.* Washington, DC: National Academy Press.

North American Association for Environmental Education. (1999). *Excellence in Environmental Education: Guidelines for Learning (K-12)*. Washington, DC: North American Association for Environmental Education.

Novak, J. D. (1998). *Learning, Creating, and Using Knowledge: Concept Maps as Facilitative Tools in Schools and Corporations*. Mahwah, NJ: Erlbaum.

Novak J. D. and Gowin, D. B. (1984). *Learning How to Learn*. Cambridge: Cambridge University Press.

Orians, G. H. (2000). Evolutionary ecology: dealing with ghosts of interactions past. Unpublished paper presented at the Ecological Society of America Annual Meeting. Snowbird, UT: August 6.

Orr, D. W. (1993). *Ecological Literacy. Education and the Transition to a Postmodern World*. Albany, NY: State University of New York Press.

Pickett, S. T. A., Kolasa, J., and Jones, C. G. (1994). *Ecological Understanding: The Nature of Theory and the Theory of Nature*. San Diego, CA: Academic Press.

Risser, P. G. (1986). Ecological literacy. *Bulletin of the Ecological Society of America*, **67**, 264–270.

Schauble, L. (1996). The development of scientific reasoning in knowledge-rich contexts. *Developmental Psychology*, **32**, 102–119.

Shu, J. (2003). The role of understanding urban ecosystems in community development. In *Understanding Urban Ecosystems: A New Frontier for Science and Education*, ed. A. R. Berkowitz, C. H. Nilon, and K. S. Hollweg. New York, NY: Springer-Verlag, pp. 39–45.

Shulman, L. S. (2002). Making differences: a table of learning. *Change*, **34**, 36–44.

Smith, G. C. (2003). Developing students' systems thinking for creating sustainable urban communities: a California model. In *Understanding Urban Ecosystems: A New Frontier for Science and Education*, ed. A. R. Berkowitz, C. H. Nilon, and K. S. Hollweg. New York, NY: Springer-Verlag, pp. 328–342.

Stevenson, R. D., Haeber, W. A., and Morris, R. A. (2003). Electronic field guides and user communities in the eco-informatics revolution. *Conservation Ecology*, **75**, 3. [Online] URL: http://www.consecol.org/vol7/iss1/art3.

Tamir, P. and Zohar, A. (2004). Anthropomorphism and teleology in reasoning about biological phenomena. *Science Education*, **75**, 57–67.

Tauber, A. I. (2001). Is biology a political science? *BioScience*, **49**, 479–486.

Wackernagle, M. and Rees, W. (1996). *Our Ecological Footprint: Reducing Human Impact on the Earth*. Gabriola Island, BC: New Society Publishers.

White, P. A. (2000). Naive analysis of food web dynamics: a study of causal judgment about complex physical systems. *Cognitive Science*, **24**, 605–650.

Zohar, A. and Ginossar, S. (1998). Lifting the taboo regarding teleology and anthropomorphism in biology education: heretical suggestions. *Science Education*, **82**, 679–697.

MILTON MCCLAREN AND BILL HAMMOND

12

Integrating education and action in environmental education

INTRODUCTION

There is a tension in environmental education between its advocates (Orr 1992; Stapp 1969a, 1969b; Wilke 1993) and those who view it as at best a distraction from the core curriculum and at worst a platform for the promulgation of radically subversive messages (Logomasini 2001; Sanera and Shaw 1996). Critics of environmental education question whether it can be seen as educational if it indoctrinates students to particular analyses of the state of the environment and advocates specific solutions to the problems identified by this selective analysis (Sanera 2002). In common with other educational initiatives developed to address social problems, many environmental education programs fall within the curriculum orientation described as social reconstruction (Eisner 1985). Social reconstruction draws its content from "pervasive and critical social problems and from the hubs of social controversy" (Eisner 1985). However, as Sauvé (1996) has noted, environmental education should address all forms of human/environment interactions, not just those that can be characterized as issues or problems.

The challenge facing those concerned with the development of an environmentally literate citizenry is to be educationally valid while accurately reflecting current debates and the state of knowledge about human–environment interactions. Environmental education could be an ideal arena for the development and practice of critical thinking, and an opportunity for students to develop an appreciation of how different subjects contribute to the development of knowledge and the solution of problems. By giving students opportunities to plan and implement actions to address real environmental problems in their communities, environmental education could provide a powerful means for the enhancement of civic literacy (Orr 1992; Stapp *et al.* 1996). However, critics of environmental education claim that it can only be educational if it presents in a balanced fashion the full spectrum of views about current environmental problems. Moreover, some have suggested that the content of environmental education programs should be restricted solely to the natural sciences and that civic and social action should not be a part of environmental education (Sanera 1995a).

In this chapter we first explore briefly the nature of education itself because all environmental education programs must be educational. We then turn to a consideration of the interdisciplinary characteristics of environmental education and explore the challenges of curriculum integration. The second section describes the different forms of action in environmental education. Finally, we give some suggestions for organizing and developing environmental education curricula with a focus on environmental action within the context of education.

THE NATURE OF EDUCATION IN ENVIRONMENTAL EDUCATION

Attributes of an educated person

Education entails the development of certain habits of mind. These include the ability to distinguish and use different forms of knowledge and to be aware of one's own learning. Philosophers of education have long argued about different forms of knowledge. Hirst (in Barrow 1984) proposed the forms as religion, philosophy, literature and the fine arts, physical science (including the life sciences), mathematics, and morals. Whether or not one accepts Hirst's categories, it seems reasonable that different forms of knowledge can be distinguished by their logical structure, concepts, and methods of establishing the validity of claims and concepts (Barrow 1981). This means that concepts that are appropriate and useful in the physical sciences may have little or no meaning or application in religion or the fine arts. As Barrow (1981) puts it: "The concepts that are peculiar to a subject have a lot to do with assessing sense and nonsense and truth and falsity in that subject."

Moreover, an educated person should be able to distinguish logically distinct questions and recognize how different disciplines conduct discourse and test claims. Barrow (1981) claims that there are only four types of questions: moral questions which are questions about what ought to happen if justice is to be served, aesthetic questions which are evaluative questions about the beauty of things, philosophical questions about the meaning and coherence of discourse, and scientific questions which are empirical questions about how things are and how they work. It is important to note that forms of knowledge are not necessarily identical to the subject categories and disciplines commonly found in the curricula of schools and universities. Biology, chemistry, and physics are separate subjects, but they address scientific questions. On the other hand, aesthetic questions appear in the arts, literature, the sciences, and mathematics.

The earliest descriptions of the field of environmental education viewed it as intrinsically inter- or multidisciplinary. The complexity of environmental questions and the environment itself, especially when humans are included, often demands the perspectives of more than one discipline. However, many multi- or interdisciplinary programs of environmental education fail to clarify or have students develop a clear appreciation of what different disciplines or forms of knowledge contribute to an understanding of environmental topics. Barrow (1981) remarks, "If you decide to teach subjects you should make sure that students come to see the application of the subjects to complex problems, and if you decide to teach through problems, you should make sure that, by the end, students have extracted the common features from various problems that go to make a discrete subject."

Besides being able to distinguish different disciplines and forms of knowledge a student should possess a number of tactics for learning in different contexts and appreciate his or her own thought processes. He or she should also be aware of the influences of their prior knowledge, beliefs, assumptions, and values on perceptions, knowledge construction, and actions. They have learned how to learn. As described here, education is clearly more than the acquisition and recall of information.

Modern learning theory holds that the mental structures through which we perceive the world profoundly affect perception (West *et al.* 1991; Bransford *et al.* 1999). These structures not only affect what we see, but also may determine what we can see. Mental structures include prior knowledge, values, beliefs, stereotypes, and assumptions. Learning entails the constant revision of existing structures and the construction of new ones. When learning is described as a constructive activity, it is mental structures that are constructed (West *et al.* 1991; Bransford *et al.* 1999). Because of differing conceptual structures, people often perceive and act upon apparently identical situations in different ways.

In our view, an educated person can stand outside their worldviews or conceptual structures and recognize how these may affect thinking and actions. Some hold that educated people should rise above emotions and values and treat the world with coldly logical detachment. We do not accept this position. We believe that an educated person should understand and appreciate his or her emotions and value them for their powers while guarding against their limitations. It is a demanding goal, worth striving for.

Education and indoctrination

A further requirement of education is that it must avoid indoctrination, defined as the planting of beliefs, generally as parts of an ideological system (Barrow 1981). Once indoctrinated, ideological systems limit how one thinks and acts. For example, a person indoctrinated to a racist ideology is very likely to behave as a racist. More important, indoctrination affects what one notices and how one interprets experience. In this way it establishes a vicious spiral, affecting not only what one knows, but also what one can know. Experience becomes a set of self-fulfilling prophecies. Doctrine replaces critical inquiry and reflection. Education is the antithesis of indoctrination. Many well-meaning educators fall into the trap of believing that indoctrination can only be performed for ideas that are false. We hold that this is not the case. All ideas, whether widely held to be good and true or not, should be open to rational and critical examination. The old Sufi saying that everything I tell you is a lie, including this, might be taken as a concise description of the required attitude.

Many environmental educators are very passionate in their care for and concern about the environment. Many are active in environmental organizations and projects. Their enthusiasms often have charismatic appeal, especially for the young. This can lead to calls to students to adopt positions and ideas uncritically without serious, thorough debate and discussion. Where teachers seek unyielding commitment to various beliefs and positions, even though they may be viewed as good, right, and noble, they are in effect indoctrinating students rather than educating them (McClaren 1989, 1993, 1997). Indoctrination does not contribute to the education of students, as worthy as the doctrines or ideologies and their associated programs of action may seem to be.

The apparent gravity of some environmental problems makes it tempting to treat environmental education as a program designed to recruit students to particular solutions, analyses, and actions. Some argue that the time for debate is over, and the time for action has arrived. Some recent "framework" documents (Government of Canada 2002) appear to claim that education, as desirable as it may be, is a slow, cumbersome, and unpredictable process so the role of environmental learning, a term used in place of education, should be to induct students into the acceptance of government policies such as sustainable development. After all, who can argue with sustainability? This sort of thinking reveals a clear lack of understanding of the nature of education and it is precisely why environmental education must remain clearly focused on both education and the environment (Jickling 1997; Sauvé 1996).

The interdisciplinary dimension of environmental education

The scientific discipline of ecology, as a branch of the biological sciences, is concerned with the study of the relationships between living organisms, including humans, and their physical, biotic and abiotic, environment. From the very beginning of ecology in the 1860s it was seen as involving disciplines other than biology, in particular physical geography, geology, meteorology, chemistry, and physics. In the 1920s and 1930s anthropology and sociology incorporated ecological ideas. Ecology has long been seen as an essential element of environmental education. Given the eclectic nature of ecology, it is little wonder that environmental education is also viewed as intrinsically inter- or multidisciplinary.

The interdisciplinary nature of environmental education notwithstanding, some argue that environmental education should focus entirely on a science-based understanding of the environment (Sanera and Shaw 1996). However, for any curriculum to have merit it must be an accurate reflection of the fields of study and methods of inquiry that are required to deal with the content covered and problems it has set itself. Narrowing the ambit of environmental education to only the natural sciences, and to a restricted list of these, would invalidate it. Instead, we must make sure that environmental education programs presented to students are not disorganized disciplinary collages, but rather demonstrate the different contributions to be made to our understanding of the environment by different forms of knowledge and disciplines.

The term integrated is commonly used to describe curricula that create opportunities for exchanges or collaborations among different subjects and disciplines. In the end, the hope is for sharing of concepts and ideas that are part of all the disciplines involved. Integration can take several forms including crossdisciplinary, multidisciplinary, pluridisciplinary, and transdisciplinary programs (Jacobs 1989). We will use the term integration here to refer to an interdisciplinary approach in which the methodology and concepts from more than one discipline are integrated by a central theme, issue, problem, topic, or experience (Jacobs 1989).

An effective environmental education requires an integrated curriculum for several reasons. First, the nature of contemporary knowledge and knowledge construction demands increasing collaboration and communication among once isolated disciplines. In addition, appropriate curriculum integration can promote a better appreciation of the way different forms of knowledge work and contribute to collaborative knowledge construction. A second argument supporting curricular integration is to reduce the curriculum fragmentation currently found in most schools and colleges. This fragmentation occurs in the scheduling of time allocations for various courses, and the lack of communication among faculty who teach different, potentially related subjects. Current learning theory views learning as a very context-sensitive activity (Bransford *et al.* 1999). Students often fail to transfer concepts and skills learned in one setting to different contexts unless definite measures are put in place to foster transfer and help them see potential linkages. By creating appropriate bridges among subjects we foster transfer of learning and greater understanding. In addition, effective integration of curricular elements can free larger blocks of time in the schedule, thereby enabling students and teachers to spend more time on larger, more complex projects. Otherwise, students and faculty alike are often the victims of a disintegrated schedule in which subjects that could be related actually compete for attention and time.

Finally, curriculum integration can help students to learn how to work with specialists in a range of fields while focusing on topics of common concern. The use of interdisciplinary teams during students' educational experiences is an appropriate form of training for many of the current realities of work beyond the classroom where project teams and work groups representing specialists from a range of fields are more and more common.

Jacobs (1989) describes two objections commonly raised against integrated programs. She refers to the first as the potpourri problem in which an integrated program takes a bit of something from one discipline and bit of something else from another. The result is a lack of depth and focus. The powers that might be contributed by the various disciplines, assuming that the topic around which they are being integrated justified it in the first place, are lost. Jacobs describes the second objection as the polarity problem. This occurs when different disciplinary specialists don't know how to collaborate, or when the common problem used as the focus for collaboration is poorly selected. Disciplinary specialists are enthusiastic about their fields. They don't like to surrender turf or time if they think students will be robbed of opportunities to become skilled and knowledgeable in their fields. This problem is exacerbated by the fact that teachers in schools and colleges traditionally do not work in teams, even within the same fields. In fact, teaching is among the most isolated of all modern forms of work. While teachers work with classes of students they seldom work with colleagues in teaching teams. Hence, when teachers are called upon to work together in interdisciplinary teams, they may lack the skills or the time needed for joint planning to make the teams effective. As a result, more than a few attempts at program integration have failed.

We believe that curriculum integration can be a powerful asset to environmental education programs, especially where programs engage in the examination of complex, multifaceted environmental processes and problems. We also believe that a foundation of core knowledge and skills in a number of relevant disciplines is essential for environmental education. Ackerman (1989) has suggested two critical questions that should be asked by anyone considering curriculum integration. First, does it make intellectual sense to integrate certain parts of the curriculum, and second, does integration make practical sense. If these questions cannot be answered in the affirmative, then it is possible that the integration of the disciplines is inappropriate and may constitute a distraction. An integrating theme that allows students from different disciplines to develop clearer and more powerful understandings of a concept such as the conservation of energy and matter will have validity for the disciplines involved. The catch, of course, is to find appropriate integrating themes or significant problems that demand application of the methods and concepts of different disciplines. Ackerman also suggests that appropriate integration of disciplines should contribute to broader educational outcomes as we have already noted above.

There are a number of practical obstacles to program integration in most educational institutions. Important among these are time, budget, and schedule (Ackerman 1989). Organizational inertia and the traditional acculturation of teachers and faculty to their work are also considerable obstacles. We will not address these issues in depth here other than to note that effective integration demands coordination and planning by instructors from different subject fields and disciplines. For teachers and faculty working on their own in closed classrooms and labs, the effort may seem to be more demanding than justified by the returns, especially in institutions that pay little attention to educational innovation or program development in assessing faculty performance. However, we agree with Ackerman (1989) to the effect that, "[w]ith its promise of unifying knowledge and modes of understanding, interdisciplinary education represents the pinnacle of curriculum development."

THE NATURE OF ACTION IN ENVIRONMENTAL EDUCATION

Two charges have been leveled against environmental education. First, environmental education is merely a façade for the recruitment of students to programs of radical environmentalism. Second, environmental education is too detached from action, too academic and abstract at a time when urgent action is needed to address major environmental problems. A clear disagreement about the nature and purposes of environmental education is apparent in these two charges. However, while action has always been a central and unique characteristic of environmental education, educators must be aware of different types of action, and the role of action projects in environmental education, if they are to escape either irrelevance or indoctrination. For example, advocates of replacing environmental education with learning for sustainability (Government of Canada 2002) or education for sustainable development (ESD) (Learning for a Sustainable Future 1993) seem to feel a strong need to recruit students to support and act on the policies of sustainability. The challenge facing those who believe there is a difference between education and indoctrination, even when the indoctrination is directed toward well-intentioned purposes or goals, is to find ways to develop a curriculum for environmental education in which students can critically consider environmental issues, examine possible courses of action, and take action as integral parts of educational experience without recruiting them to selected causes or policies and indoctrinating them to particular ideologies.

Let us begin by stating a few of our premises. First, it must be assumed that all the learning within a program will be directed toward educational ends; in other words, what is learned about, through, and from action should contribute to developing the habits of mind and attributes of character we have described above. Second, as was the case in our examination of curriculum integration and interdisciplinary programs, we believe that the primary consideration in attempting any inclusion of action in a curriculum should be whether or not it is educational to do so. Third, we believe there is a distinction to be made among learning about action, learning through action, and learning from action, a distinction we examine more fully below. We believe that it is vitally important for action projects to be very carefully chosen and selected in terms of their appropriateness for the curricular purposes we wish to attain, the students who are to be involved, the communities in which the actions will be situated, and their overall practicality given the time and resources available. However, this does not mean that projects should be selected by faculty and merely implemented by students.

We believe it is vitally important for projects to be developed through a process of collaboration and negotiation that includes students and faculty and possibly members of the larger community as well. There are several reasons for this. First, where teachers select projects for classes or students, they become open to charges of recruiting students to pet projects or causes, even if they have no intention of doing so. Second, and more important, one of the key things to be learned about action is how to select and define appropriate projects. Thus, students should be involved in considering projects, evaluating their feasibility and appropriateness, and negotiating any disagreements among those who are involved. A corollary to this is the principle that if students should be given the right to select and evaluate projects, they should also be given the right to opt out of such projects, either as individuals or groups. That right should be respected, and preserved against peer pressures. Moreover, it should be quite possible to have different projects being undertaken by students in the same class who either have different priorities or diverse points of view on a common case or issue. If we are to avoid even the subtle indoctrination of students, we must make every effort to not only respect diversity, but also empower it. We should note, however, that teachers should be active participants in the process of selecting, defining, and negotiating action projects. Teachers should employ vetoes if, in their professional judgment, a project is inappropriate. This is acceptable as long as the students and other participants understand the roles and responsibilities of teachers from the outset.

Learning about, through, and from action

Hammond (1997a, 1997b) has proposed a three-part approach to action studies: learning *about* action, learning *through* action, and learning *from* action. We believe it is very important for instructors and curriculum developers to be clear about the differences among these three perspectives because each involves different instructional models and learning environments.

Learning about action

Learning about action entails students learning skills and strategies removed from actual action projects, although hypothetical or simulated projects and cases can be used. This often involves teaching students the skills of action, the history of action projects, and providing them with examples of action projects as models. Students can study cases. They can read accounts and histories of the work of natural and social scientists, activists, naturalists, and conservationists (Knapp 1993). They can also study principles of ecological management and apply them to simulations, role-plays, mock hearings about environmental issues, or selected classroom-based problems. These are all useful approaches. Hungerford *et al.* (1992) have developed an extensive set of skill modules "designed to teach students how to investigate and evaluate science related social issues." The modules are intended to foster the acquisition of a number of useful skills. They can serve as a foundation for learning from and through action.

Learning about action may inspire some students to develop projects on their own challenges outside the school program or through the extra curriculum. However, in many cases students who learn knowledge and skills in schools fail to appreciate their application to situations outside the classroom or examination hall and do not transfer their knowledge and skills to new settings. Instructional programs in emergency medicine, flight training, and military combat are notable in not assuming transfer of learning from classrooms to real world situations. To facilitate transfer students are engaged in highly realistic simulations, case studies, and challenge exercises that attempt to replicate the venues in which they will be expected to apply their training. Hence, while it is important and useful to learn about action, no one should assume that students would automatically transfer knowledge and skills from the classroom to real life situations.

Many business schools, medical faculties, and engineering schools have introduced instructional approaches intended to enhance higher order thinking skills and foster students' abilities to work in collaborative teams. These tactics may be broadly clustered under the terms problem-based learning or case studies. Effective problems or cases share several attributes (Duch 1997). For example, cases should be authentic and engage students in the examination of real world problems. While being authentic, a good case should also have enough complexity to challenge students' interest and motivate them to seek understanding. A good case should also require students to apply processes of reasoning and to make and justify decisions based on facts, information, and logic or rationalization. A good case will require students to explain assumptions, decide what information is relevant, and select appropriate methodologies. Additionally, good cases often demand student cooperation and collaboration or at least better solutions are likely to be reached through collaborative knowledge construction. A good case should also be open-ended, without having one best answer or solution. While it should be possible to apply previously learned knowledge and skills to the case, there should not be a "right" answer to be found at the back of a textbook. Fifth, a good case should also have controversial or contentious elements that will bring forth diverse opinions within the problem-solving team. This will require teams to learn and apply skills in negotiation and conflict resolution. Finally, a good case should be relevant to the concepts and skills of the disciplines in which the PBL approach is used. Effective problem-based learning is not simply a motivational device for students; it should teach them to think through and with the tools of the disciplines.

Good cases are often framed as narratives or stories, creating situations or scenarios. The Skill Development Modules developed by Hungerford *et al.* (1992) provide examples of the approach. In some forms of PBL students are placed in full-blown simulations in which they assume roles and are required to stay within their role throughout the process. This method requires students to look at a problem from unfamiliar personal, ideological, scientific, social or cultural perspectives, thereby developing an appreciation of the views and thoughts of others.

Learning through action

Learning through action entails students getting directly involved in action projects outside classroom-based cases and examples. Although learning through action shares many attributes with problem-based learning, the important difference is that, in learning through action, the action is not simulated. Students work in the real world on projects that have tangible outcomes other than reports, presentations, proposed solutions, or decisions. A second difference is that while cases or problems in problem-based learning are often designed and/or selected by faculty, learning through action projects are usually the result of guided negotiation in which the students play central roles in identifying and defining action projects.

Projects can range from specific, short-term, single-issue projects to those requiring sustained engagement over long periods of time. However, in all types of action projects there are rich opportunities for students to both learn and apply skills and knowledge and gain experiences that can contribute to attitude change and shape character. When students are involved in selecting, planning, implementing, and evaluating effective projects, there are opportunities to develop an enhanced sense of personal competence and overcome the syndrome of powerlessness. The importance of engaging in action is that there is often information to be gained through action that is simply not available in classroom presentations or exercises, or even from the best simulations, no matter how well designed and realistic they are. In practice the best-laid plans do indeed often go astray. Vitally important learning opportunities are presented when students have to change plans in midstream. Furthermore, real world projects have real world consequences, consequences that have direct relevance to the students involved in the projects. The knowledge and skills that students may have gained through programs that focus on learning about action are extremely useful here.

Wilke (1993) has categorized the primary methods through which a person or group may engage in environmental action as: persuasion, consumerism, political action, ecomanagement, and legal action. Hammond (1997a, 1997b) classifies action projects in three categories. In Level 1 projects students engage in the design, development, and implementation of action projects that produce products or have tangible outcomes. Typical examples might include building a nature trail, cleaning up a stream or vacant lot, implementing garbageless lunches in schools, collecting aluminum to sell in order to buy critical habitat, planting butterfly gardens, or building and installing nest boxes for bird species or bats. This is the level at which the vast majority of teachers and students usually engage in environmental projects.

Level 2 projects require an additional set of implementation skills and are characterized by the design, development, and implementation of ongoing, long-term, continuing projects. Examples include establishing and maintaining a school recycling program, permanently maintaining a wildlife management area, establishing and maintaining a fish hatchery, or developing and implementing a school or district-wide "green" purchasing and operations plan. School-based long-term environmental monitoring programs like GLOBE (no date) also function at this level.

Level 3 projects require even more sophisticated skills than those at Level 2 and are the least common in educational settings. They are characterized by the design and implementation of changes in policies, regulations, or laws. Examples can be found where students have drafted legislation and lobbied legislators to pass new federal, state, or local laws. They range from proposing laws declaring a state insect or reptile, to habitat protection ordinances, changes to school board policies regarding the use of post-consumer recycled paper products or energy consumption, convincing land developers to set aside parklands or conservation areas, getting land acquisition projects placed on the ballots of *ad valorem* tax referenda, and successfully filing suits against environmental offenders (these include the political, legal, and often all other action categories used by Wilke (1993)).

Clearly, each level demands different amounts of engagement, skill, knowledge, and resources, including time. Level 3 projects are the most likely to be contentious or controversial and are therefore the most likely to involve classes, teachers, or schools in controversy and criticism. Orr (1992) claims that environmental education is inextricably political, not in the sense of conventional partisan politics, but in the sense that politics is the process by which communities discuss values, make choices from among options, and allocate priorities. Education in democratic societies requires that students understand politics, not as abstract theory or as the fierce partisan struggle among modern mass-media-driven political parties, but as a fundamental and practical requirement of life in democracies.

Learning through action is aided by a number of instructional strategies and learning environments. We should note that, before any teacher, or team of instructors, undertakes to work with students in learning through action projects, it is essential to establish a learning community that is characterized by respect for diversity, trust, and willingness to engage in collaborative learning and to contribute to the learning of all members of the class. Because projects in this category are less commonly undertaken as solo works, students and faculty alike will benefit from understanding cooperative learning, teamwork and group process, conflict negotiation and resolution, values clarification, and communications. Good research skills are also essential. Once again, skills learned from programs that focus on learning about action will be invaluable. In our view, learning through action presents the ultimate opportunity for the collaborative construction of knowledge and for the social, symbolic, and physical distribution of intelligence among students (Perkins 1992).

Learning from action

Learning from action comes into play when students review projects, cases undertaken in learning about action programs, or their own direct experiences in community projects. It involves assessment of project outcomes and processes in regard to what can be learned for future projects and for their lives as citizens in a community. This approach can be seen as a phase of both learning about action and learning through action programs, because it is often to be found as a concluding element of experiences in those. However, we believe that it has so much significance that it justifies separate consideration. The claim has been made that engagement in projects may result in more active and effective civic participation by students. If this claim is to be realized in practice, it requires that students not only study action, or even take action, but also reflect on the significance of actions for themselves and their communities. For example, many students have been engaged in clean-up projects on local school grounds. While the clean-up is a useful action, it is not likely to lead to longer-term changes in student participation in civic government or politics or to major changes in student lifestyles unless teachers and students systematically reflect on the reasons why the clean-up was necessary in the first place, and consider whether or not they have addressed the root causes, and what changes might be needed to affect cures as opposed to short-term fixes.

Learning from action requires a set of skills on the part of both students and teachers that is different from those applied or gained in learning about or through action. A powerful approach is the keeping of working journals and field logs by teachers and students. It is also very important for students to present and communicate the results of projects. Communicating to those who were not involved is a powerful way of consolidating learning. Teachers must develop the ability to conduct effective debriefing sessions with individual students and with collaborative groups involved in projects. Debriefing is of critical importance to this approach. It can also be helpful to invite community members who represent a range of perspectives to serve as critical reviewers of action projects during the debriefing phase.

Action projects in practice

In school programs it often appears that only learning about action is present, or is stressed to the exclusion of learning through or from action. It is comfortable for teachers and students to stay within the safe confines of the classroom and to treat environmental action as a concept abstracted from implementation and context. Surely, classroom environments can support more than learning about action. Although there are wonderful examples and role models that can inspire and inform students (Knapp 1993), these alone are not generally sufficient to encourage and enable students to make the commitments of time and effort required to successfully take action outside of school. If students decide to embark on action projects there is a critical need for the support, experience, and wise counsel of teachers and community members. This can be possible when students' early experiences with action occur in the ambit of sound educational programs in schools and colleges. Otherwise, it is as if we had taught students the theory of flight and the history of aviation, introduced them to the pioneers of flight, and then let them jump into the cockpit of an aircraft to take their first solos. To make the example more telling, we then might blame the students either for failing, performing badly, or being reluctant to take off at all.

Many teachers and faculty members have good reasons for not wanting to undertake environmental action projects with classes. While some are obviously concerned about lack of support from colleagues, administrators, or the community, or fear criticism directed at them for radicalizing students and recruiting them to selected causes, others feel a lack of knowledge and skill concerning how to initiate and manage action projects, especially those that move beyond studying about action and begin to move into projects that entail learning through action. In this context it is very useful to have guidelines and templates for the management of action projects in educational contexts.

FRAMEWORKS AND PROCESSES FOR INTEGRATING EDUCATION AND ACTION

Mitchell and Stapp (1996) and Ramsey *et al.* (1989) have developed a number of very useful frameworks for environmental action in schools. Mitchell and Stapp's (1996) work was the inspiration for projects like GREEN and GLOBE that continue today to involve schools in many nations, while frameworks and process models developed by Hungerford and his associates have had an important influence on environmental education standards in the US. Recent thinking about the nature of action research and experiential learning justifies an attempt to provide some new suggestions for thinking about and planning for learning about and through action, particularly in the context of interdisciplinary curricula. Figures 12.1 and 12.2 are offered not as tightly defined frameworks of how action projects must be undertaken but as suggestions for planning curricula. These suggestions are for those educators who are taking their first steps toward integrating action studies as an element of environmental education.

A framework for learning about action

A framework for a case- or problem-based approach to learning about action is provided in Figure 12.1. Although intended to be used in the context of an interdisciplinary, integrated curriculum, the framework could also be used as a discussion guide in a course with a unidisciplinary organization. In that case the outline would serve as a guide to possible connections to other relevant fields of knowledge or disciplines. However, the actual involvement of students and faculty in a real world problem would be lacking.

There are several critical points in this framework. First, the selection of the case (Step 1) is critical. We have described some criteria for case selection previously in this chapter (Duch 1997). In most case- or problem-based programs instructors play an important role in identifying cases that are likely to be appropriate and have educational power. However, there is no reason why student teams cannot identify cases and refine or negotiate those proposals with the assistance and advice of the instructor or instructional team. This will be particularly feasible where classes have gained experience with problem-based learning by being guided from simpler examples to more complex, less-well-defined, and controversial ones. A great deal of learning in this approach is derived from students struggling with messy problems that may be difficult to define and that lack one best answer.

A second critical point in the process diagram is Step 10. This step entails students making sense of the information they have gathered while examining the case from the various disciplinary perspectives (Steps 3–9). One of the most important things that can be learned here is that information does not automatically turn into actions. Gathering

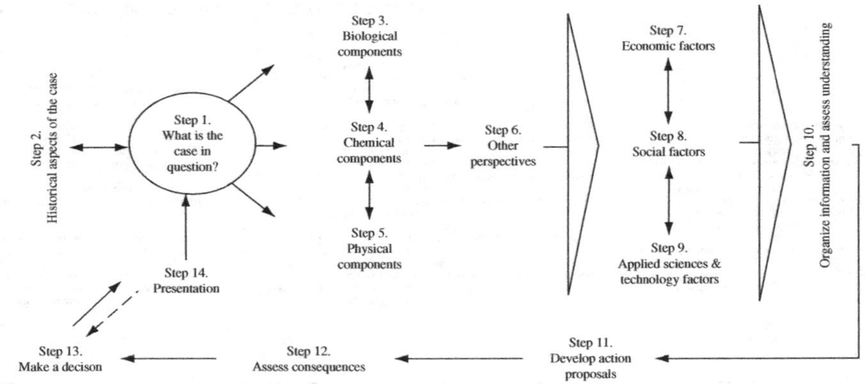

Legend

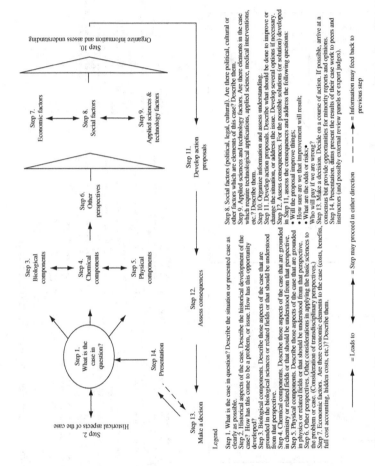

Step 1. What is the case in question? Describe the situation or presented case as clearly as possible.
Step 2. Historical aspects of the case. Describe the historical development of the case? How has this come to be a problem, or issue. How has this opportunity developed?
Step 3. Biological components. Describe those aspects of the case that are grounded in the biological sciences or related fields or that should be understood from that perspective.
Step 4. Chemical components. Describe those aspects of the case that are grounded in chemistry or related fields or that should be understood from that perspective.
Step 5. Physical components. Describe those aspects of the case that are grounded in physics or related fields or that should be understood from that perspective.
Step 6. Other perspectives. Other considerations in applying the basic sciences to the problem or case. (Consideration of transdisciplinary perspectives.)
Step 7. Economic factors. Are there economic elements to the case (costs, benefits, full cost accounting, hidden costs, etc.)? Describe them.
Step 8. Social factors (political, legal, cultural). Are there political, cultural or other factors which are elements of this case? Describe them.
Step 9. Applied sciences and technology factors. Are there elements in the case which require technological applications, applied science, medical interventions, etc.? Describe them.
Step 10. Organize information and assess understanding.
Step 11. Develop action proposals. Describe what should be done to improve or change the situation, or address the issue. Develop several options if necessary.
Step 12. Assess consequences. For the possible solutions (or solution) developed in Step 11, assess the consequences and address the following questions:
• Will the proposal improve things;
• How sure are we that improvement will result;
• What are the odds or risks;•
Who will pay if we are wrong?
Step 13. Make a decision. Decide on a course of action. If possible, arrive at a consensus but provide opportunities for minority reports and opinions.
Step 14. Presentation. Teams present the results of their case work to peers and instructors (and possibly external review panels or expert judges).

→ = Leads to ↔ = Step may proceed in either direction - - → = Information may feed back to previous step

Figure 12.1. A process for learning about action projects using a case- or problem-based learning approach.

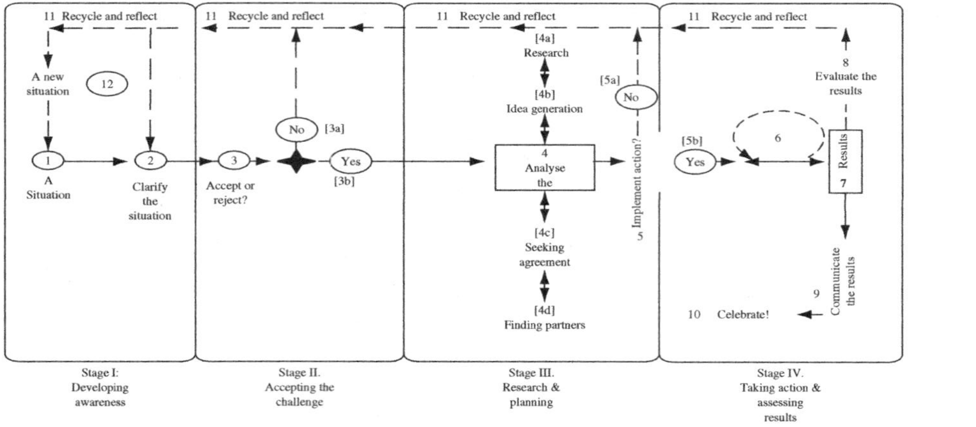

Stage I. Developing awareness
1. A situation. A situation becomes apparent. Gather first impressions; develop awareness of the situation.
2. Clarify the situation.
Has the situation been described the situation as clearly as possible?
Is the situation: a problem; an issue; an opportunity?

Stage II. Accepting the challenge
3. Accept or reject. Acceptance or rejection of the situation.
Do we want to try to do something about the situation?
[3a] No.
- We don't care enough to want to do something.
- The situation is too big for us to handle.
- We don't have the skill.
- We don't have the time...

[3b] Yes.
- We are prepared to accept responsibility.
- We want to try to do something about the situation.

Stage III. Research and planning
4. Analyse the situation.
What do we need to do at this point? What must we learn to do in order to be likely to succeed?
[4a] Research.
Do we need more information and research before we can develop a plan or implement action?
[4b] Idea generation.
Do we need to generate new ideas about what might be done?
[4c] Seeking agreement.
How can we agree about what should be done? How can we find consensus; resolve conflict?
[4d] Finding partners.
Who can help us with this project? How can we engage their assistance?
5. Implement action?
Should we implement action?
[5a] No
We should not implement action.
[5b] Yes.

Stage IV. Taking action and assessing results
6. Action & learning in the course of the action project.
This step may loop back several times.
7. Results.
An outcome or outcomes of the action project.
8. Evaluate the results.
- What has happened so far?
- Were the project goals accomplished?
- What still needs to be done?
- What have we learned?

9. Communicate the results.
10. Celebrate the accomplishments.
11. Recycle and reflect.
12. A new situation (or perhaps the original situation revisited with new insights and skills).

Figure 12.2. A process for learning through action.

more information will not necessarily help in selecting an action proposal from among several options (Step 11). At some point a decision must be made, remembering that decisions not to act, or to defer action while further research is undertaken, are still actions and also have consequences. There is usually no such thing as a certain, single, absolutely best solution to complex real world problems.

In Step 13 on the diagram students, having weighed the consequences in Step 12, should make a decision. If possible, that decision should reflect a consensus because the exercise of seeking consensus is an important aspect of group process and can be an important collateral outcome of this method. However, if a consensus cannot be found, especially bearing in mind previous comments about indoctrination, then minority or dissenting options should be encouraged and given time in the final presentation of results (Step 14). Students attach greater significance to the presentation of decisions if their reports are juried by critical outside reviewers who may be community members with experience in the issues, or experts in various subject fields.

While problem-based learning and case studies do not require that students actually implement their action proposals, it is possible to add zest to this process by injecting "wild cards" or treating the process as a simulation with role plays. In a role-play approach each team is given the case or scenario (and cases are often presented in story format). Each team member is then given a particular role to assume (member of the town council, environmental activist, representative of a local business with a stake in the issue, etc.). Roles can be quite detailed and descriptive and the players may be told not to share some aspects of their role requirements with the other members of the team but to play them authentically. Further, as the teams are working, they may suddenly be confronted by unexpected new information, perhaps delivered by an actor who assumes a particular role (a process server delivering a subpoena; a reporter from a local paper wanting a story or wanting a reaction to a "leaked" document). These approaches can add considerable reality and relevance, to say nothing of interest and fun, to the case method.

A framework for learning through action

Figure 12.2 provides a process for managing environmental education programs with real world action projects as elements of the learning experience. We offer this framework not as the only way to view the process of learning through action but as a possible approach. If we consider learning through action programs as falling within the broader category of action research, then the literature on alternative frameworks is very large. Lewin (1948) characterized the process of action research as having three major components: planning, including reconnaissance of the problem, taking action, and assessment of and reflection on the action. These processes were not seen as being arranged in a linear series, but rather in a series of feedback loops in which each part of the process could feed both back and forward to the other stages in an iterative process. Another way of viewing the process is simply learning by doing. In this sense, the approach formalizes the process of building on experience by making it explicit and documenting experiential learning, thereby enhancing opportunities for improving and developing the project.

The four major stages of the process in Figure 12.2 are similar to those proposed by Lewin (1948), except that we have isolated the acceptance of the challenge or project as a distinct stage. In the context of educational programs, we believe that it is critical for students to think clearly before deciding to take on action projects of any kind. Classes may drift into action or be carried away by momentary enthusiasms or emotional appeals unless they think critically about the implications of the project for themselves, their colleagues, and communities. Hence, step 3 on the diagram represents a critical choice point at which a team or class (or perhaps an individual) explicitly decides to proceed with action work. Note that this is not a final decision and that external review can be extremely helpful here. At point 5 on the diagram the action team has another opportunity to decide whether to implement action. At both points there are loops back that can move the students back to points in the process for further consideration, or take them back to the starting point. While some may be tempted to view such backward movement as a failure, we prefer to see it as an important opportunity for learning from action through reflection while presenting opportunities for students to recognize the need for better information or further skill development. As is implied by Point 11 on the figure, recycling and reflection may even lead to the selection of a new action project, possibly one that is more appropriate.

At Stage II it is important for the group to consider what kind of problem or situation is involved. Note that the situation may be a problem or it may be an issue, a distinction made by Hungerford *et al.* (1992) wherein environmental problems are viewed as situations involving humans, the environment, and relationships between the two in which the well-being of humans is threatened. It is, of course, a very anthropocentric definition but it is a useful starting point for considering the difference between a problem and an issue. According to Hungerford *et al.* (1992) an issue is a problem about which two or more parties cannot agree on solutions or about the seriousness or nature of the problem. As Hungerford and his coworkers note, many environmental problems remain unaddressed because people cannot agree about how to solve them.

Note, however, that a situation having potential for action may be neither a problem nor an issue, but an opportunity. Opportunities are situations in which the well-being of humans or the integrity of ecosystems is not threatened and which may not be controversial, while offering potential for the application of knowledge and skill, development of awareness and understanding, exercise of creativity, or the enjoyment of life. All human–environment interactions should not be seen as problematic or controversial situations (Sauvé 1996; Hammond 1998). The construction of a nature trail or bird-viewing platform in a conservancy or the publication of nature art created by local community members are projects that develop opportunities rather than solve problems. There can be environmental actions that are joyful, playful, inventive, and celebratory. They should not be missed or dismissed as possible projects.

The research and planning stage (Stage III) of Figure 12.2 shows a number of steps that might be taken under the broad heading of analysis. They include research (4a) (of various kinds depending on the nature of the action project and the disciplines seen as being able to contribute), idea generation (4b), negotiation (seeking agreement) (4c), and developing partnerships and collaborations with stakeholders and community members (4d). In all of these, skills and knowledge that may have been gained and practiced in regular subjects or in learning about action programs will be very useful. Learning through action projects are powerful ways of applying skills and knowledge, deepening understanding, and developing enhanced performance. We should note that the figure presents the four stages of the overall process as being of approximately equal scale. There is no need for this to be so. In some action projects it can take months, or even years, for awareness to develop while in others that stage may be quite compressed and the stage of taking action is lengthy.

Step 9 involves communicating what has been done and learned to a wider audience and making a project public. Students can derive very important knowledge and experiences from this aspect of action work. Sometimes people get so wrapped up in their projects that they don't take time to communicate with others. As a result, they may fail to build networks of support or, more important, miss opportunities for powerful learning experiences through which they have to clarify their own thinking in order for others to understand. Step 8 on Figure 12.2 involves evaluation of the results through summative and formative evaluations (Scriven 1967). Formative evaluations are ongoing throughout the recycle and reflection phases of the process, and are extended across all four phases shown on the diagram.

It should be noted that the framework diagram suffers from the limitations of all two-dimensional representations of complex processes that occur through time and that are iterative. While the double-headed arrows on the figure indicate potential feedbacks and feed forward processes, in reality, the stages are constantly in flux as a consequence of changing conditions and new information. Nevertheless, we believe the diagram is a useful tool for thinking about learning through action projects in environmental education.

CONCLUSION

While action research can employ conventional experimental scientific methodology in components of the overall research work, it differs from experimental science in that the researcher is not detached from the situation, all variables are not controlled, and the subjects may also be participants in the research project. In action research, research is not distinct from implementation. However, this characteristic of action research also gives rise to a possible concern for its inclusion in educational programs. When students and faculty invest time, effort, and resources in the implementation of an action, they may no longer be able to view its consequences with critical detachment and separate inquiry and understanding from advocacy. This problem may be at least partially addressed by involving external reviewers from outside the specific action project group, either from other classes or faculty in the school or from the community at large, in the evaluation of all project stages, proposals, and outcomes.

Action research is by definition not value-neutral: the action researcher seeks not only to understand a system but also to make a constructive or positive change, an improvement in the system, a clearly value-laden stance. We prefer to regard action research not as a distinct methodology as much as a synthesis of a number of methodologies, including experimental science, qualitative methodologies, narrative and historical or ethnographic approaches, and soft systems science. We believe it is a mistake to set action research rigidly apart from more conventional quantitative and experimental scientific methods. An action research project that involves improving an urban stream might need the rigor of research in ecological restoration and stream ecology as part of an overall suite of approaches required to understand and improve the situation. We believe that where students move beyond the classroom to engage in action projects, they will almost inevitably find themselves engaged in some form of action research, regardless of which disciplines form the starting points for research and planning.

As Joyce and Weil (1972) have noted, all models of teaching are both nurturing and instructive. Environmental education programs present unique opportunities for the development of the habits associated with the educated mind. They can also nurture the qualities we claim to be essential for citizenship in democracies. Appropriately designed, educationally sound action projects that entail learning through action, as well as learning about and from it, are powerful assets for both instruction and the nurture of environmentally educated citizens. With clear thinking about the nature of education, it is possible to reconcile and integrate education with environmental action, without recruiting students to causes and indoctrinating them to single points of view. In fact, we contend that to do otherwise is to nurture a sense of incompetence, powerlessness, and a view of the disciplines and forms of knowledge as irrelevant and abstracted from life.

REFERENCES

Ackerman, D. (1989). Intellectual and practical criteria for successful curriculum integration. In *Interdisciplinary Curriculum: Design and Implementation,* ed. H. H. Jacobs. Alexandria, VA: Association for Supervision and Curriculum Development, pp. 25–37.

Barrow, R. (1981). *The Philosophy of Schooling.* Brighton, UK: Wheatsheaf.

(1984). *Giving Teaching Back to Teachers: A Critical Introduction to Curriculum Theory.* Brighton, UK: Wheatsheaf Books/London, ON: The Althouse Press.

Bransford, J. D., Brown, A. L., and Cocking, R. R. (1999). *How People Learn. Brain, Mind, Experience, and School.* Washington, DC: National Academy Press.

Duch, B. (1997, Feb 20). *PBL. (University of Delaware).* [Online] URL: http://www.udel.edu/pbl/cte/spr96-phys.html (retrieved November 29, 2002).

Eisner, E. W. (1985). *The Educational Imagination,* 2nd edn. New York, NY: Macmillan.

GLOBE. (n. d.). GLOBE is Supported by NASA, NSF, EPA and the US Dept. of State. GLOBE is a Worldwide Hands-on, Primary and Secondary School-based Education and Science Program. [Online] URL: http://www.globe.gov/globe_flash.html (retrieved, December 9, 2002).

Government of Canada. (2002). *A Framework for Environmental Learning and Sustainability in Canada.* Ottawa: Government of Canada.

Hammond, W. F. (1997a). Learning about action, learning from action, learning through action. *Clearing*, **99**, 7–11.

(1997b). Educating for action. *Green Teacher*, **50**, 6–12.

(1998). *Earth is Not a Problem.* Unpublished Ph.D. Dissertation. Burnaby, BC: Simon Fraser University.

Hungerford, H. R., Litherland, R. A., Peyton, B. R., Ramsey, J. M., and Volk, T. L. (1992). *Investigating and Evaluating Environmental Issues and Actions: Skill Development Modules.* Champaign, IL: Stipes Publishing Co.

Jacobs, H. H. (1989). *Interdisciplinary Curriculum: Design and Implementation.* Alexandria, VA: Association for Supervision and Curriculum Development.

Joyce, B. and Weil, M. (1972). *Models of Teaching.* Englewood Cliffs, NJ: Prentice-Hall.

Knapp, C. E. (1993). *Environmental Heroes and Heroines: An Instructional Unit in Earth Values and Ethics.* Oregon, IL: The Environmental Education Association of Illinois.

Learning for a Sustainable Future. (1993). *Developing a Cooperative Framework for Sustainable Development Education.* Ottawa, ON: National Round Table on the Environment and the Economy.

Lewin, K. (1948). Resolving social conflicts. In *Selected Papers on Group Dynamics [1935–1946]*, ed. G. W. Lewis. New York, NY: Harper.

Logomasini, A. (2001, July 6). *Environmentalist Tampering with Our Children.* The Competitive Enterprise Institute. [On line] URL: https://secure.cei.org/gencon/019,02083.cfm (retrieved Nov. 26, 2002).

McClaren, M. (1989). Environmental literacy: a critical element of a liberal education for the 21st century. *Education Manitoba*, **17** (1), 2–12.

(1993). Education not ideology. *Green Teacher*, **35**, 17–18.

(1997). Reflections on "Alternatives to national standards in environmental education: process-based quality assessment." *Canadian Journal of Environmental Education*, **2**, 35–47.

Mitchell, M. K. and Stapp, W. B. (1996). *Field Manual for Water Quality Monitoring: An Environmental Education Program for Schools.* Dubuque, IA: Kendall/Hunt Publishing Company.

Orr, D. W. (1992). *Ecological Literacy: Education and the Transition to a Postmodern World.* Albany, NY: State University of New York Press.

Perkins, D. (1992). *Smart Schools. From Training Memories to Educating Minds.* New York, NY: The Free Press/Macmillan.

Ramsey, J. M., Hungerford, H. R., and Volk, T. L. (1989b). A technique for analyzing environmental issues. *Journal of Environmental Education*, **21**(1), 26–30.

Sanera, M. (1995a). *Kids and the Environment: Taught to be Warriors, Worriers. Center for Environmental Education Research*, Center for Environmental Education Research Publication No. 2394112. Claremont, CA: The Claremont Institute.

(2002). *The EPA's Support for Biased and Politicized Environmental Education.* The Competitive Enterprise Institute. [Online] URL: https://secure.cei.org/gencon/004,02412.cfm (retrieved November 29, 2002). February 27.

Sanera, M. and Shaw, J. S. (1996). *Facts Not Fear: A Parent's Guide to Teaching Children about the Environment.* Washington, DC: Regnery Publishing.

Sauvé, L. (1996). Environmental education and sustainable development: a further appraisal. *Canadian Journal of Environmental Education*, **1**, 7–34.

Scriven, M. (1967). The methodology of evaluation. In *Perspectives on Curriculum Evaluation*, ed. R. Tyler, R. Gagné, and M. Scriven. Chicago, IL: Rand McNally.

Stapp, W. B. (1969a). The concept of environmental education. *American Biology Teacher*, **32**, 14–15.

(1969b) The concept of environmental education. *Journal of Environmental Education*, **1**(1), 30–31.
Stapp, W. B., Wals, A. E. J., and Stankorb, S. L. (1996). *Environmental Education for Empowerment: Action Research and Community Problem Solving.* Dubuque, IA: Kendall/Hunt Publishing Company.
West, C. K., Farmer, J. A., and Wolff, P. M. (1991). *Instructional Design: Implications from Cognitive Science.* Englewood Cliffs, NJ: Prentice-Hall.
Wilke, R. J., ed. (1993). *Environmental Education. Teacher Resource Handbook.* Millwood, NY: Kraus International.

MARIANNE E. KRASNY AND RICK BONNEY

13

Environmental education through citizen science and participatory action research

INTRODUCTION

Ecologists and environmental scientists from many universities in the United States and Canada are engaged in environmental education through Citizen Science, Student-Scientist Partnerships, and Participatory Action Research projects. Citizen Science and Student-Scientist Partnership programs are unique in that they offer opportunities for scientists to work with K-12 students, teachers, and other community members while furthering their own research (Cohen 1997; TERC 1997). In these programs, scientists who have a need for data from a large geographic area, yet do not have the resources to collect such data themselves, enlist volunteers (Citizen Science) and students in classrooms (Student-Scientist Partnerships) to help. Many of these programs are designed to monitor broad-scale and long-term environmental phenomena, such as changes in animal populations, water quality, or global climate. Examples include Project FeederWatch, Monarch Watch, the Terrestrial Salamander Monitoring Program, and the international Global Learning to Benefit the Environment (GLOBE) program. Designing Citizen Science and Student - Scientist Partnership programs entails balancing the scientists' need for data with the educational community's interest in improving student understanding and attitudes about science and the environment.

Participatory Action Research (PAR) presents an alternate framework for how scientists can become involved with community members. PAR combines research with education and action - it attempts to engage local people in defining and conducting research, with the goal of not only collecting data, but also educating community members about their current situation and engaging them in actions to improve local conditions (Gaventa 1998; Chambers 1999; Freudenberger

1999). PAR was originally conceived by social scientists and activists who felt that researchers working in poor communities sometimes exploited local people by extracting data for their own purposes with little consideration of how the people might benefit. Community members involved in PAR learn and become empowered through the data collection process and through subsequent community action. Because PAR often uses "hands-on" methods to gather information, including participatory mapping, diagram drawing, and transect walks (an interview conducted while walking across a site), it is useful in situations where there are language differences between researchers and local people. Although originally designed for work with adults, PAR approaches also have been used in youth environmental science education programs (Doyle and Krasny 2002; Mordock and Krasny 2001).

In addition to deciding on a Citizen Science, PAR, or other approach to guide the design of educational activities, designers of environmental education programs must develop a process for working with the educators who will teach the activities. Efforts to involve educators vary from using a top-down "train-the-trainer" approach to a more participatory "learning community" process. Whereas the concept of learning communities has been used in undergraduate education and other settings, the definition used in university agriculture extension can perhaps best be adapted for use in other university outreach programs including those focusing on environmental education. Defined as "settings for information exchange and innovation among individuals with varying knowledge," agricultural learning communities draw on the expertise of both farmers and scientists to build understanding of growing practices and resource management (Millar and Curtis 1999; Jordan *et al.* 2000). Similar to PAR, learning communities value local or farmer knowledge, in addition to the expertise of scientists. Farmers in an agricultural learning community experiment with ways to adapt new practices to the social and environmental conditions on their individual farms, and then share the results of their experiments with other farmers and scientists, thus furthering an overall understanding of new practices. Such communities can have a positive effect on the adoption of new practices (Roling and Wagemakers 1998; Ison and Russell 1999; Wuest *et al.* 1999) and on communication among researchers, extension agents, and community members (Gerber 1992). Applied to an environmental education context, learning communities involve educators in conducting local programs, each of which may be viewed as an informal research project whose goal is to determine best educational practices (Krasny and Doyle 2002; Krasny and Lee 2002). Thus, the "local" knowledge of community educators is combined with the "expertise" of university program developers and scientists. In another parallel with farmer learning communities, educators may play an important role in communicating with other educators and thus in program dissemination.

This chapter addresses some of the challenges in using Citizen Science and PAR approaches to develop environmental education programs that include science education, scientific data collection, and environmental activism goals. We also consider questions related to balancing science and education: How does the program engage scientists given other demands on their time? How does the program work with educators and youth to ensure that the program meets their needs? We focus on two examples of university outreach programs. First, we describe the history of the pioneering Citizen Science efforts at the Cornell Laboratory of Ornithology (CLO) (www.birds.cornell.edu.citsci), where Citizen Science is now integral to the organization's scientific and educational mission. Next, we discuss the Cornell University Garden Mosaics program (www.gardenmosaics.org), which can be viewed as an experiment in combining Citizen Science data collection with PAR community activism. Furthermore, the Garden Mosaics example illustrates a learning community approach to working with educators to define the program goals and activities. Following the descriptions of the CLO and Garden Mosaics programs, we discuss key issues related to designing Citizen Science programs, drawing on the literature and the two case study programs.

CORNELL LAB OF ORNITHOLOGY CITIZEN SCIENCE PROGRAMS

Program goals and development

The field of ornithology has a centuries-old tradition of amateurs making observations, and thus contributing to our understanding of changing bird populations. For example, in 1900 the Christmas Bird Count started as an amateur bird counting project by the National Audubon Society (National Audubon Society 2003). In the early 1960s, CLO embarked on an effort to formalize scientific collaboration with amateur birdwatchers through its Nest Record Program. Building on the success of this initial project, scientists at CLO and Bird Studies

Canada launched Project FeederWatch in 1987, with the goal of using data collected by amateur birdwatchers to study the dynamics of winter bird populations and to determine factors affecting population changes. FeederWatch was the first of what were to become many current CLO Citizen Science initiatives.

Project FeederWatch was pioneering in its attempt to engage a large public audience in collecting standardized data. While CLO scientists knew that thousands - some said millions - of individuals put up feeders and watched the birds that visited, they did not know whether a large number of birders would be willing to gather and report data in a uniform manner. They also were unsure whether data collected by volunteers could be used to uncover regional and continental population patterns. To facilitate data collection by non-scientists, CLO staff devised a "count protocol" that was simple, easy to understand, and would yield useful data even if observers spent only a small amount of time watching feeders. They then began recruiting participants - called "FeederWatchers" - through advertisements, direct mail, and notices in CLO and other newsletters.

More than 3,000 participants signed up during the 1987–88 winter, each paying a minimal fee to cover the cost of project materials. When the data began arriving in the spring, CLO scientists determined that the information was, in fact, useful for studying winter bird populations, and that if the project continued, the data were likely to reveal important population trends. However, the scientists also realized that if the project were to grow, they would need to pay significant attention to the needs of the participants. For example, participants called or wrote with many questions, both about the protocols and about bird biology. All those questions required responses. Participants also wanted quick feedback about the results of their data-collection efforts. At the same time, CLO realized that the participants' interest created tremendous unanticipated opportunities. FeederWatch could be important not only for gathering scientific data, but also for educating the public about bird biology, ecology, and the scientific process. CLO staff could help participants understand the answers to questions such as: How and why was a particular protocol developed? How did the scientists analyze the data? How did scientists use the results? What further questions about bird populations did the data raise?

To seize the educational opportunities, while at the same time addressing issues related to retaining and supporting volunteers, the scientists turned to CLO's education and publications staff.

Working together they devised a long-range plan for FeederWatch that included both scientific and educational goals. The educators developed materials that clearly explained the significance of the project and its potential findings, and "research kits" with clear instructions, data forms, tally sheets, and identification posters. They then field tested the protocols with amateurs and worked with the scientists to refine the procedures where necessary. The educators also assumed responsibility for recruiting and providing support to participants, interacting with those who had questions, ideas, and problems as they carried out the project. In addition, they produced a newsletter to provide project results in a timely and engaging manner. This division of labor freed CLO scientists to concentrate on analyzing the huge quantities of FeederWatch data. Furthermore, the scientists had time to submit project results to refereed journals, thus disseminating the new information to the scientific community.

Fifteen years later, the concept that CLO dubbed Citizen Science has proven successful as a tool for both research and education. CLO has engaged more than 100 000 citizens of all ages and backgrounds in Citizen Science projects, which in addition to Project FeederWatch include Classroom FeederWatch, The Birdhouse Network, Urban Bird Studies (including Project PigeonWatch), Birds in Forested Landscapes, the Great Backyard Bird Count, and eBird, among others (see www.birds.cornell.edu/citsci/ for an overview). The projects vary in difficulty of data collection protocols, from counting birds at a feeder (Project FeederWatch) to selecting count sites, describing site habitats, and using timed bird counts and playback of recorded songs and calls (Birds in Forested Landscapes). Furthermore, many other organizations have adapted the Citizen Science model for projects ranging from tracking monarch butterfly migrations to monitoring atmospheric conditions (Cohen 1997; TERC 1997; GLOBE 2003).

Challenges

Building a comprehensive Citizen Science program that meets the needs of scientists, educators, and the public has presented several challenges. These can be grouped into five general areas: (1) balancing science and education; (2) ensuring data quality; (3) forming partnerships to enhance educational goals; (4) evaluating impacts; and (5) building institutional support for Citizen Science.

Balancing science and education

CLO educators and scientists are committed to developing Citizen Science projects in which volunteers investigate authentic research questions - questions for which scientists truly desire to know the answers. From a scientific point of view this objective is obvious. However, from an educational standpoint, engaging lay people in real research raises questions about what kinds of scientific questions are appropriate for various audiences and how the research promotes educational goals.

CLO Citizen Science projects generally originate when scientists have questions that can be answered only with large numbers of observers working over a wide geographic area for an extended period of time. Example questions and the projects designed to answer them include the following: How do intraspecific clutch sizes differ with latitude and climate? (The Birdhouse Network). How is bird breeding success affected by the quantity and quality of available habitat? (Birds in Forested Landscapes). How quickly does disease spread through wild bird populations? (House Finch Disease Survey). The data collected by volunteers in these projects have led to numerous refereed journal publications focusing on how bird populations change in distribution over time and space (Wells *et al.* 1998; Hochachka *et al.* 1999); how the breeding success of certain species is affected by environmental change (Rosenberg *et al.* 1999); how emerging infectious diseases spread through wild animal populations (Hochachka and Dhondt 2000; Hartup *et al.* 2001); and how acid rain may be affecting bird populations (Hames *et al.* 2002).

The original Citizen Science participants (birdwatchers) were excellent data collectors and thus readily contributed to CLO's research goals. However, they were not necessarily the audience most in need of science or environmental education. Demographic studies of FeederWatchers showed that participants were primarily middle class, well educated, and already knowledgeable about science issues (Bonney and Dhondt 1997). To diversify its audiences, CLO educators and their scientist colleagues developed several Citizen Science projects where the educational goals assumed priority over the research goals. For example, in an effort to reach urban youth, CLO educators initiated Project PigeonWatch and worked with scientists to define the specific research questions: Why do city pigeons exist in so many colors? What color mate does a pigeon choose? Subsequently,

CLO created new and adapted existing projects for a more diverse audience (e.g., Birds in the City, Dove Detectives, and Crows Count).

CLO educators also noted that none of the original protocols involved actual experimentation on the part of the participants - an important type of research for young people to experience. Thus, the educators designed several experimental protocols, including the National Seed Preference Test in which volunteers used experiments to determine whether birds at ground feeders prefer black-oil sunflower, white proso millet, or red milo seeds. At the time this project began, most information on birdseed preferences was anecdotal or based on a few geographically isolated studies and suggested that birds do not like red milo. Interestingly, Project FeederWatch participants from the southwestern United States and elsewhere had reported that their birds loved red milo. Indeed, results from the Seed Preference Test showed that many ground-feeding birds do eat milo.

Finally, CLO educators wanted to help all project participants, even those with significant science knowledge, evolve from being data collectors to true citizen scientists. This entailed participants learning to ask and answer their own questions in addition to collecting data for studies designed by scientists. Thus, the educators developed materials to help all participants become involved in the scientific process from start to finish. For example, the Classroom FeederWatch curriculum, teacher workshops, and website include suggested research questions and extension activities, and guide teachers and students through the process of designing and conducting original research. Students can submit the results of their research to CLO for publication in its student research journal *Classroom BirdScope*. Similar efforts are being made to enhance inquiry-based learning in CLO's informal science education programs.

Although the results of these educator-initiated programs have not been published in scientific journals, they have proven useful to scientists and others. For example, results from Project PigeonWatch were presented at meetings of the American Ornithologists' Union (LaBranche *et al.* 2000) and results from the Seed Preference Test have been used by the birdseed industry. CLO scientists also have reported the findings of these projects in the CLO newsletter *BirdScope*.

Ensuring data quality

Data quality is an important issue for any Citizen Science project that intends to publish the results. To ensure data quality, CLO scientists and educators address three major areas: data collection protocol, data editing and analysis, and observer training.

At CLO, each project has developed data collection procedures that volunteers can readily follow and that yield accurate results. For example, consider the problem of tracking birds that visit feeders. How can a participant document the number of birds that visit a feeder on any given day? Simply counting birds over the course of the day does not work, because few if any volunteers can watch their feeders all day long. Besides, the same birds might make several visits, and thus be counted multiple times.

CLO scientists involved in Project FeederWatch solved this problem by devising a sampling procedure that is largely independent of the amount of time an observer is able to spend. Each time observers look at a feeder, they record the number of individuals of any one species present within the feeder area, and report the largest of those numbers in any one day as the daily count. For example, a FeederWatcher who observed a feeder three times on a Sunday might see one Northern Cardinal the first time, three the second time, and two the third time. They would report three cardinals on the data form, the largest number of individuals that were observed at any one time. Stated another way, this observer is telling CLO scientists that at least three cardinals were present in the yard on a given day. While this sampling procedure does not yield precise bird counts, it does produce data that are comparable among observers, describe relative abundance, and can be used to detect population trends.

Not all volunteer observers are expert birders and some certainly make mistakes. Errors are minimized through careful editing during data entry and before data analysis. For example, CLO's electronic data forms are "smart" meaning they include only species likely to be seen in the observer's area, and allow only counts that are within preselected limits. If an observer enters erroneous data, either through a misidentification or a keyboard mistype, a friendly message asks the observer to double-check the entry. If the observer truly has seen an unusual species or an unusually large number of individuals, they can override the error message as long as they send corroborating details. CLO staff constantly adjust the data editing procedure to reflect new findings about bird numbers and distributions. Before any quantitative analyses, outliers are flagged and double-checked, sometimes by contacting the observer. Obviously spurious data are eliminated. Furthermore, CLO scientists are experienced at evaluating large "messy" data sets by using statistical methods that search for patterns. In a project that includes hundreds or thousands of observers, patterns are fairly easy to determine even if some data are not completely accurate.

To further ensure data quality CLO offers training opportunities, particularly for projects that target teachers and urban youth who may not be skilled birdwatchers. Educational materials such as bird identification posters, CDs of bird vocalizations, and online field guides are included on project web sites and in research kits, and are available to all participants, including those who do not attend trainings.

Forming partnerships to enhance educational goals

CLO was able to form long-term partnerships with numerous science education organizations, including the Museum of Science and Industry in Tampa, the Franklin Institute in Philadelphia, the Urban Ecology Center in Milwaukee, and the Urban Ecology Institute in Boston, through a collaboration with the American Association for the Advancement of Science. Educators from these organizations have worked closely with CLO staff to pilot project materials and to provide feedback on whether youth understand the reasons for the project, use the materials productively, and follow the protocols easily. This has resulted in modification of the projects to meet the needs of the educators and youth participants while at the same time maintaining scientific integrity. For instance, the piloting of Birds in the City resulted in changes in the length and width of transects that participants follow as they count birds along city streets.

At the same time, partnering with other organizations presented challenges. Groups such as Scouts or Boys and Girls Clubs generally have their own objectives, agendas, projects, and ways of presenting materials. Simply preparing a leader's guide for an existing CLO project, such as The Birdhouse Network, and then expecting a given youth group to adopt the project and use it appropriately was generally doomed to failure. Nowhere was this challenge greater than in classrooms, where teachers are under ever-increasing pressure to prepare students for standardized tests.

The key to successfully adapting and disseminating projects has been to work closely with educators to develop materials that they can meld into their own programs. For example, CLO has adapted FeederWatch for use in middle schools by developing a standards- and inquiry-based curriculum called Classroom FeederWatch. This curriculum was developed over three years with extensive input from more than 100 middle school teachers across North America.

The teachers helped to develop, pilot, and field test the curriculum, and later helped to train additional teachers in its use. As a result of teacher input, the curriculum and accompanying slides and posters cover the required subject matter (e.g., diversity, adaptation, and graphing skills), thus increasing the likelihood that the project will be used in classrooms. CLO also developed a Classroom FeederWatch website to facilitate learning and data entry and retrieval by students (http://birds.cornell.edu/cfw/).

Evaluating impacts

CLO collaborates with faculty and graduate students in the Cornell University Education and Communication departments to conduct project evaluations. In most CLO Citizen Science projects participants appear to enhance their research skills and understanding. For example, students use CLO's online databases to formulate and answer questions and they conduct web searches to find related data. Furthermore, students perceive their Classroom FeederWatch experiences - both collecting data and designing their own projects - as real scientific investigations (Trumbull *et al.* 1998, 1999a, 1999b; Chakane and Trumbull 2000). Finally, the number of submissions to *Classroom BirdScope* that represent authentic inquiry has increased each year of the project.

Similarly, it appears that the process of participating in the Seed Preference Test contributed to the participants' thinking about biology and the scientific process (Trumbull *et al.* 2000). The evaluation of this project used approximately 800 unsolicited letters from participants, which were scored for evidence of hypothesis formation and testing and other learning. The results indicate that participants often made observations about the ecology of their feeding sites or about animal behavior, and some formulated careful hypotheses or made suggestions for modification of the experimental design.

Results from pre- and post-tests of participants in The Birdhouse Network showed an increase in knowledge of the biology and habitat needs of cavity-nesting birds, but failed to demonstrate a change in the understanding of the scientific process, or in attitudes toward science and the environment. The lack of change in these areas may indicate a need for including more explicit information about the scientific process in the research kits and may reflect the fact that many participants already had positive attitudes toward the environment. A pre/post assessment of Project PigeonWatch participants revealed that they changed their views about scientists. When asked to draw a scientist before they began participating in the project, most children drew stereotypical scientists (lab coat, messy hair, male). In contrast, after they had observed and collected data about pigeons, many of the children drew scientists who looked less stereotyped and more like themselves (Lewenstein 2001).

Building institutional support for Citizen Science

Citizen Science at CLO is a serious undertaking, requiring both commitment and resources. In addition to having a leadership that supports research–education collaborations, CLO's unique infrastructure and talented staff make such projects possible. Unlike other university departments, CLO is a membership organization. Its membership, communications, and marketing staff are well positioned to assist in the large tasks of recruiting participants and maintaining participant databases. Its membership fees can be used to help support Citizen Science projects. CLO also has a strong information technology staff that design and maintain both the relational databases that hold the project data, and the interactive data forms that participants use to submit, retrieve, explore, and interpret data.

Furthermore, CLO has been successful in obtaining large grants to develop its research and education efforts through the National Science Foundation and several private foundations (e.g., Wallace Genetic Foundation). Some organizations that have used specific project data, including the US Forest Service and the Environmental Protection Agency, also have provided significant financial support. However, most of the grants are limited to a project's startup stage; obtaining funds to keep projects operational has been more difficult. In addition to support from membership fees, participant fees have been used to maintain projects. For example, both Project FeederWatch (with more than 15,000 participants who pay $15 USD per year) and Classroom FeederWatch (whose materials sell for $99 USD with a $20 USD annual renewal) are close to self-sustaining.

Finally, CLO not only has strong research and education programs, but the organization as a whole has a common mission and the scientific staff are intensely devoted to education and outreach. In fact, commitment to education and outreach is one of the criteria used in hiring new scientists, just as an understanding of science is essential for the education staff. The success of CLO's Citizen Science program depends on this commitment and collaboration between scientific and educational staff.

In conclusion, the scientist – educator collaboration at CLO runs smoothly for three reasons. First, both parties benefit, either from the data that are collected or from the unique educational experiences Citizen Science provides. Secondly, leadership at CLO supports both educational and scientific goals and has created a culture where both are important. In fact, Citizen Science is the *only* way that many CLO scientists are able to conduct their research, and scientists often do not even view Citizen Science as outreach but rather as a normal part of conducting research. Finally, the educators and scientists have developed an agreement on how to divide the labor, with the education and publications staff supporting the participant network, and the scientific staff analyzing and publishing the data.

GARDEN MOSAICS: PARTICIPATORY ACTION RESEARCH AND CITIZEN SCIENCE

Program development and activities

Unlike many of the CLO Citizen Science programs, Garden Mosaics did not originate with a research question generated by a scientist. Instead, Garden Mosaics began with the realization that community gardens are unique and underutilized sites for environmental science education programs in cities. This uniqueness stems from the fact that they often are the only green space in densely populated neighborhoods where people can enjoy nature, socialize with neighbors, relax, and participate in cultural and educational events, as well as grow fresh food. They also bring together people representing a mix of ages and immigrant and minority cultures, and the gardeners' planting practices often draw from their unique cultural traditions. Furthermore, some of the growing practices originated in areas where access to commercial fertilizers and pesticides was limited, and thus farmers depended on more sustainable ways of adding nutrients (such as composting and intercropping), conserving water (e.g. raised beds, mounds, and furrows), and controlling pests (e.g. use of marigolds to repel nematodes, use of soap solutions in place of commercial pesticides). The name Garden Mosaics reflects this "mosaic of cultures," as well as the "mosaic of planting practices" that the gardeners create.

Drawing from both Citizen Science and PAR approaches, the Garden Mosaics youth program includes *Investigations* in which youth gather information about the neighborhood, gardens and gardeners;

and *Action Projects* to benefit the garden and community. In the first investigation, *Neighborhood Exploration*, youth use aerial photographs, maps, and a walk through their neighborhood to see where, besides the garden, residents can find fresh food and places for socializing, relaxing, and cultural and educational events. They then produce a neighborhood collage using maps, photos, and other results of their exploration. The second investigation, *Garden Hike*, involves a walk through a community garden with a knowledgeable gardener, during which the youth interview the gardener and make observations about what is growing, what structures are present, and what activities take place in the garden. Youth report this information to the *Community Garden Inventory* database on the Garden Mosaics website. In the third investigation, youth develop a *Gardener Story* for the website, detailing a gardener's planting practices, where the practices originated, and any planting tips the gardener may suggest.

In addition to sharing the results of the Garden Hike and Gardener Story online similar to other Citizen Science programs, the youth use what they learn during the investigations about the problems the gardeners face to define an Action Project. These projects provide opportunities for the youth to do something meaningful for the gardeners and their community. The projects vary from designing a new garden to cooking a meal at a soup kitchen to conducting an educational activity in the garden.

Program goals and challenges

The Garden Mosaics activities combine multiple goals for the youth participants including: (1) conducting research; (2) participating in a Citizen Science project; (3) learning science content; (4) conducting an environmental action; and (5) forming partnerships with elders. In addition, Garden Mosaics has created a "learning community" of educators and scientists, which plays a key role in defining the program goals and activities, and also provides professional development opportunities for educators. Each of these is discussed below, along with some of the successes and challenges that Garden Mosaics has faced in meeting these multiple goals.

Goals for youth

In Garden Mosaics, the youth research is largely descriptive; it entails documenting ethnic and sustainable planting practices and the gardens' contributions to the social and cultural fabric of the community. Garden Mosaics originally attempted to use PAR and related Participatory Rural Appraisal (Freudenberger 1999) methods for the youth research because these approaches were developed for use in communities similar to the US urban immigrant and minority neighborhoods where Garden Mosaics is being implemented (i.e., where people are often poor and lack political power, and where there are language differences between the researchers and community members). In addition, Garden Mosaics was originally committed to a participatory process of involving community members in defining the research questions. However, several issues arose that led us away from a more participatory approach and toward the development of standard research questions and investigations as described above. First, it became apparent that educators and students needed more guidance to reach the science education learning goals than was being provided by such an open-ended approach. Furthermore, community members were not necessarily interested in becoming co-researchers (cf. Saldivar-Tanaka and Krasny 2005). Finally, the program staff saw the value of collaboration with non-profit groups and scientists that is possible only if the questions are the same at multiple sites. Thus, similar to CLO, Garden Mosaics starts with structured investigations, and then provides opportunities for students to define their own research questions based on what they have learned and on the interests of the gardeners.

Although the youth's research is not strictly PAR in that they start with questions that are defined by the program rather than by community members, the youth apply several PAR principles and approaches in their investigations. First, the youth conduct investigations not only to document food-growing practices, but also to learn about problems that the gardeners face and to define an Action Project they later conduct to help the gardeners. Thus, the youth's research is not simply to "extract" information from the gardeners, but also to "give back" to the community. Second, some of the methods that the youth use in their research are drawn from Participatory Rural Appraisal. For example, the youth collect information using "transect walks," or mobile interviews conducted while walking across a site with a gardener. Initially, we used many of the more "hands-on" Participatory Rural Appraisal activities such as participatory mapping and diagramming, but the youth had difficulty leading these activities with the gardeners. A third PAR principle, honoring the "traditional" knowledge of community members in addition to taking scientific measurements, is also integral to the youth's research.

Youth participating in Garden Mosaics create several databases for use by the gardening and scientific community. The first database, developed in collaboration with the American Community Gardening Association, is an inventory of community gardens, including membership, history, and the amenities provided to surrounding neighborhoods. The second database documents ethnic and sustainable gardening practices of individual gardeners. Furthermore, an urban weed database has been added to the Garden Mosaics website to document the diversity of weeds and weed management practices in urban gardens. Similar to other Citizen Science projects, the Garden Mosaics databases produce information that is useful to scientists. For example, the weed data will inform a research program on organic means of urban weed management. The Gardener Stories could be used by agricultural scientists to investigate new practices and ethnic crops for commercial agriculture.

However, the Garden Mosaics databases differ from other Citizen Science efforts in several ways. First, Garden Mosaics research provides information not only for use by scientists, but also by non-profit groups and individuals. For example, the inventory of community gardens is designed to be used by organizations seeking to build a case for the importance of these gardens in maintaining the social and cultural fabric of neighborhoods. Many individual gardeners seeking more sustainable growing methods, as well as school children and community members interested in the stories of elders, may be interested in the "traditional knowledge" that is being posted to the website and that otherwise might be lost as older gardeners become less active. Second, Garden Mosaics Citizen Science includes not only observational data, but also information garnered from the gardeners about planting practices and the relation of those practices to culture. Finally, Garden Mosaics is unique in that it seeks answers to social science questions (What is the role of the garden in the social and cultural life of the community? How do the growing practices relate to the gardeners' country of origin?), in addition to biological and physical sciences questions (What is growing? How is it being grown? What are the soil characteristics?).

Through their interviews and other research in the gardens, youth in Garden Mosaics are exposed to many biological, physical, and social sciences concepts related to urban gardening. However, the gardeners may not be able to explain the scientific principles behind their gardening practices and in some cases even may convey inaccurate information. The Garden Mosaics *Science Pages* are designed to enhance the science learning that occurs through youth research in the gardens. The pages focus on a range of topics, including the biology of plants found in the gardens, soil testing, the science behind gardening practices (e.g., biological control, inter-planting), and how to use aerial photographs and topographic maps. Educators can use the Science Pages at "teachable moments" during the youth's investigations in the gardens or in preparing for or processing the investigations with the youth. For example, if the youth learn from the gardeners about liming soil, they can conduct the activities on the pH Science Page once they have returned to their community center. If they find an unusual plant such as alache or papalo they can read about its biology on the appropriate Science Page. Each Science Page includes a short color version for the web and a printable version for youth that covers the same information as the color version plus learning activities and highlights from current scientific research; a printable version for educators that includes teaching tips is also available for some Science Pages.

In addition to providing a means for youth to "give back" to their community, the Action Projects are important in motivating older youth to participate in the investigations and other science learning activities. Although younger youth readily engage in the science learning and investigative activities, older youth often demand a rationale beyond that of providing data for scientists and non-profit groups or of enhancing their science understanding. The Garden Mosaics website includes an Action Projects database where youth can share what they have accomplished through this aspect of the program.

Educators cited youth developing positive relationships with and learning from the gardeners as one of the most important outcomes of Garden Mosaics (Krasny and Doyle 2002). Elder gardeners greatly appreciate the interest youth show in their knowledge about gardening and often express the desire to involve more youth in the garden. Furthermore, by incorporating elders into an environmental education program, Garden Mosaics provides positive role models for youth from *within* their community, something that many environmental education programs fail to do. A number of the youth–elder partnerships have continued beyond the formal Garden Mosaics activities.

Learning communities: a participatory approach to program development

Consistent with its participatory philosophy, Garden Mosaics engages cooperating educators in setting program goals and defining the youth activities. The program uses a learning community model developed in the context of sustainable agriculture and natural resources management in its work with educators. Educators from non-profit greening groups, Cooperative Extension, and universities from 11 US cities and program directors from Cornell University, Cornell Cooperative Extension–New York City, and the Institute of Ecosystem Studies constitute the program learning community. In addition to playing a pivotal role in defining program direction, members of the learning community have provided regional leadership for Garden Mosaics programs and have developed new professional skills.

For example, the educators discovered that the original research questions and Participatory Rural Appraisal protocols were not suitable for youth conducting research in community garden settings. Through a series of discussions among the program directors and educators, Garden Mosaics adopted investigations, which are engaging for youth and provide valuable information for use by scientists, gardeners, and non-profit groups. The educators also insisted that Action Projects be a fundamental aspect of the program in addition to science inquiry, both as a means to motivate youth and to give back to the community. Throughout the learning community discussions, the educators and project directors have worked together to balance the science learning, Citizen Science, and action goals of the program.

Pre- and mid-program written surveys were used to determine the success of the learning community model in meeting its goals of professional development for the educators. Most educators appreciated the professional development and networking opportunities and felt they learned through their experiences piloting the program (Thompson 2002). About half the educators reported an increase in their knowledge about gardens and gardening, and in their understanding of youth interests, motivation, and learning. The more experienced educators felt they were lending their expertise to Garden Mosaics, and didn't expect to grow professionally from their participation.

Initially the educators expressed frustration with the learning community approach, largely stemming from wanting to be handed something that is more standardized and easier to implement (e.g., Project Learning Tree). However, developing an urban youth program that combines elements of Citizen Science and PAR is challenging, and the educators' involvement has been essential. Furthermore, as the program has progressed, the educators have increasingly expressed their strong commitment to its long-term success and sustainability.

In conclusion, Garden Mosaics has been successful in creating a highly participatory process of program development. Furthermore, it has created a program model that balances several key components of environmental education: traditional knowledge and scientific knowledge; inquiry and action; social sciences and biological sciences; and building positive, mutually beneficial relations among youth and elders in their community and among youth and scientists.

A DISCUSSION OF CITIZEN SCIENCE

In 1997, Robert Tinker laid out a vision for the future of Student-Scientist Partnerships that could also be applied to Citizen Science. He suggested that these programs could grow to have the following impacts:

- Every student could have an authentic science experience at least once while in school;
- Every school could have at least one teacher-scientist involved in original studies, changing the school culture and infusing authentic science throughout a school;
- Scientists planning research projects could routinely use schools for large-scale data collection.

While not all students have become engaged in Student–Scientist Partnerships or Citizen Science, a lot has been learned about how these programs work. By examining factors related to research discipline, data quality, inquiry learning and environmental action, scientist involvement, and educator support, we can gain insights about designing Citizen Science programs in environmental education.

Research discipline

Even though Tinker (1997) suggests Student–Scientist Partnerships can be conducted in any number of environmental, health, sociocultural, and educational disciplines, the majority of Citizen Science and Student–Scientist Partnership programs focus on monitoring animal populations or physical parameters related to climate, soils, and water. Monitoring studies lend themselves to Citizen Science because scientists often do not have the human resources to collect the data themselves and thus the success of their research depends on volunteers. Furthermore, students and the public often readily grasp the research questions and procedures involved in monitoring studies. A limited number of Citizen Science projects have engaged participants in experimental, as opposed to monitoring research (e.g., the Seed

Preference Test, coverboard experiments of the Terrestrial Salamander Monitoring Program (http://www.mp1-pwrc.usgs.gov/sally/)). In addition, for those disciplines where it is not possible to engage diverse audiences in fieldwork, scientists have provided samples to schools for analysis. For example, the Mars Exploration and Education Program provides images of the planet Mars for students to examine (Barstow and Diarra 1997), GLOBE and other Student–Scientist Partnerships have engaged students in groundtruthing (i.e., collection of field data to verify remote sensing data) (Becker *et al.* 1998; Brooks 2000), and the Paleontological Research Institution sends fossil/soil samples to schools for children to analyze (Harnik and Ross 2003b; Ross *et al.* 2003).

Having defined questions that are accessible to a lay audience, Citizen Science programs often adapt existing protocols so that they are simple to perform and entail only inexpensive equipment and supplies. For example, the GLOBE protocols can readily be conducted on school grounds and require only minimal supplies (GLOBE provides the expensive equipment when it is required). However, GLOBE teachers have experienced challenges in trying to keep up with the year-round sampling schedule required for some of the protocols (SRI 2002). The protocols for the Terrestrial Salamander Monitoring Program also pose challenges for school groups. For example, frog monitoring requires participants to locate a pond or other sampling site, learn to distinguish the calls of different frog species, time their sampling during the few weeks of the year when frogs are actively mating, and conduct the work at night, which may be dangerous and inconvenient for a youth group.

Most Citizen Science programs are in disciplines where there is a long history of volunteer involvement, such as ornithology, paleontology (Harnik and Ross 2003a), astronomy (Barstow and Diarra 1997), and atmospheric sciences (GLOBE 2003). In other fields, such as herpetology, there is a smaller pool of amateur observers to draw from, which may partly explain why such programs have not garnered as much interest as those in ornithology (e.g., the Terrestrial Salamander Monitoring Program, North American Amphibian Monitoring Program: http://www.mp2-pwrc.usgs.gov/naamp/, Frogwatch USA: http://www.nwf.org/frogwatchUSA/).

Similar to birdwatching and stargazing, gardening and sportfishing are popular leisure activities yet there has been relatively little Citizen Science activity in these areas (see, http://www.hort.cornell. edu/vlb/ and Brooks *et al.* 1999 for exceptions). This is in spite of the fact that there is a long history of farmer contributions to agricultural science (e.g., farmers developed many of our horticultural and food plants, and often collaborate with scientists to conduct field trials), and that anglers collaborate with scientists through angler diary programs (Pollock *et al.* 1994). In both horticulture and sportfishing, universities have developed educational programs that engage volunteers in teaching rather than in research (e.g., Master Gardeners, Master Anglers).

Citizen Science programs also benefit when the study subjects capture the public's attention. In addition to the ornithology examples discussed above, Citizen Science projects in astronomy and paleontology have capitalized on people's interest and sense of wonder related to stars and fossils. The Paleontological Research Institution mastodon project provides an example of media-generated interest in a Citizen Science project. Within six months of newspapers reporting the discovery of a complete mastodon skeleton, nearly 2000 schools had requested soil samples from around the skeleton for their students to process. However, only 30 percent of schools that requested samples actually analyzed and returned the fossils, indicating that the initial media-generated enthusiasm for the project may have waned (Ross *et al.* 2003).

Data quality

Despite the concerns about data quality that often are raised in relation to Citizen Science programs, a number of studies have demonstrated that given proper training and appropriate protocols, children as young as elementary school pupils can collect high quality data (Rock and Lauten 1996; Becker *et al.* 1998; Brooks 2000; Lawless and Rock 1998). For example, in the Devonian Seas Project conducted by the Paleontological Research Institution, students were able to collect good data on a limited number of fossils, but had difficulties identifying rare fossils. The scientists adjusted the protocols so that students worked only on fossils that they were likely to identify correctly and thus improved data quality (Harnik and Ross 2003b).

Inquiry learning and environmental action

GLOBE evaluations have shown that students improved their understanding of content related to the specific protocols they conducted and of environmental processes. GLOBE students also demonstrated their ability to take measurements, conduct observations, and interpret data, and were able to make more inferences regarding the environment and ecosystems compared to non-GLOBE students (SRI 1997, 1998, 2000, 2002).

However, students who engage in the research process do not necessarily develop an understanding of the way in which research is conducted (Lederman 1998). In a study evaluating the ForestWatch Student Scientist Partnership, Moss *et al.* (1998) found that most students' conceptual understanding of research did not evolve over the course of the year. For example, the students had uninformed notions of scientific questioning and viewed data collection as simply following a set of proscribed steps. The authors suggest that this was because the students' time was spent mostly on data collection, and that they did not have the opportunity to analyze their data, communicate their findings, or pose research questions on their own. Lederman (1998) suggests that students need structured opportunities to critically reflect on the research process.

Incorporating opportunities for students to develop research questions may be important not only from a motivational standpoint (students will be more interested in their own questions), but also from a science education perspective (being able to develop research questions is an important skill). However, students need guidance in learning how to develop good research questions, and participation in a Citizen Science project necessitates researching a common question. In the CLO, Garden Mosaics, and some other Citizen Science programs, students are encouraged to develop and investigate their own questions after first conducting research using a standardized protocol.

Another important educational component is students' understanding of the relationship of their research to larger social issues. The National Science Education Standards call for students to conduct research within the context of their community (NRC 1996) and experience suggests that providing students with a broader context for their research motivates them to study science. However, some youth may require more than a relevant issue to get them engaged in research; additional motivation may be provided by opportunities to engage in actions to benefit the community.

Many of the critical thinking skills entailed in designing an environmental action project are similar to those involved in research and are also a focus of the environmental education standards (NAAEE 1999). These include questioning assumptions, understanding cause and effect relationships, considering alternative explanations, and debating critically within the community. In an evaluation of Earth Force, a program that involves data collection and environmental action, Melchior and

Bailis (2003) found that students who undertook community actions demonstrated enhanced environmental citizenship skills, attitudes, and knowledge, whereas the outcomes were less positive for students who did not complete the community action project.

Scientist involvement

Collaborations with scientists are an integral part of Citizen Science and Student–Scientist Partnerships and can add to the overall experience for participants. Students are motivated by feeling they are part of an authentic research project and by having the opportunity to meet and communicate with scientists (SRI 2002; Ross *et al.* 2003).

The question of how scientists benefit from their involvement in Student–Scientist Partnerships was addressed by GLOBE. Three years after the start of GLOBE, four out of eight GLOBE scientists used student-collected data to validate their models or to compare their data with previously collected data, three presented results at GLOBE sponsored meetings, and two presented their work at research conferences and published in educational (not science) professional journals (SRI 1998). In 2003, GLOBE chief scientist wrote:

> GLOBE data have been more used for conference presentations to date. There have also been education journal articles. Scientists are using the data. The best concrete example to date is not a paper but the inclusion of data from 59 GLOBE schools in the database at the National Climatic Data Center for use in their assessment of weather events. A results paper from the Hydrology team has been accepted for publication and is in the process of final preparation. GLOBE data have also been used in several Masters theses at the University of New Hampshire. Several additional papers are in preparation. (D. Butler, personal communication, 6/10/03)

Interestingly, although scientists may initially be motivated to participate in a Citizen Science or Student–Scientist Partnership project because of their need for data, they may experience additional unexpected benefits. The three-year GLOBE evaluation reported that five of the eight scientists felt involvement in GLOBE had caused them to view their field of study in a more integrative or interdisciplinary way. Furthermore, seven scientists participated in Internet communications with schools and collaborated with schools on investigations, papers, or conference presentations (SRI 1998). Similarly, CLO and other Cornell scientists, who initially became involved in Citizen Science because of their need for data or in response to pressure from funding agencies, now find they enjoy involving students in their research. Krasny (1999) found that scientists have numerous motivations for incorporating high school students into their research labs, including wanting to "pay back" for similar research experiences they had when younger, being able to provide opportunities for their graduate students to mentor younger students, and having a more diverse group in their laboratory.

Educator support

Most of the literature on Citizen Science and Student–Scientist Partnerships focuses on the balance between the interests of scientists and students, rather than on the needs of educators. This is in spite of the fact that teachers and/or non-formal educators are a critical element in these programs and that their needs and interests differ from those of students. For example, educators leading students in research projects often need support in learning about the research process and in learning how to balance structure and allow for intellectual curiosity among students.

The year-six GLOBE evaluation examined why teachers who had been to GLOBE workshops differed in their implementation rates (SRI 2002). They found that personal mentoring by experienced GLOBE teachers was more important than other kinds of support (e.g., newsletters, follow-up trainings). Similarly, our work with learning communities in non-formal educational settings has pointed to the importance of support networks of educators, as well as of educators' interest in learning and in professional development (Krasny and Doyle 2002; Krasny and Lee 2002). These results point to the potential of a learning community approach to Citizen Science programs, in which educators are part of an informal "research/program development" team that plays a key role in setting program direction, designing implementation methods, and conducting the evaluation, with the goal of furthering our understanding of best educational practices. Integral to such an approach is a person who is assigned to support the learning community of educators, through site visits, ongoing communications, encouraging educators to participate in professional meetings, sharing program results, and other means.

CONCLUSION

The goals of environmental education vary widely, with different programs emphasizing various aspects of environmental knowledge, attitudes, and behaviors. For example, natural history programs seek to increase students' knowledge of nature, whereas other programs have more of a focus on environmental issues (e.g., pollution, overpopulation). Projects that attempt to change behaviors may range from being proscriptive in regards to a single behavior (e.g., promoting recycling, conserving energy), to promoting critical thinking and decision-making often related to community action (e.g., participating in local environmental policy decisions). Whereas some environmental educators promote holistic approaches that combine values, spirituality, the humanities, and science, others have focused more narrowly on science process and content and have attempted to relate the goals of environmental education to those of the national science education reform movement (NRC 1996; Krasny *et al.* 2001; 2002; Trautmann *et al.* 2001, 2003; Carlsen *et al.* 2004).

Citizen Science programs that are well designed, taking into account the interests of both the volunteer participants and the scientists, and that are well supported institutionally and financially have demonstrated their potential to address the science literacy goals of environmental education as well as to contribute to scientific research. Furthermore, they represent an approach that is supported by the scientific community, and thus does not lend itself to some of the charges of bias and incorrect information being leveled against environmental education.

On the other hand, Citizen Science does not directly address some of the behavioral goals of environmental education, particularly those that focus on changing personal behaviors (e.g., saving energy). Rather, Citizen Science defines behavior change in terms of critical thinking, and it is hoped that through engaging in such thinking during their research, participants will be better able to analyze information about environmental issues and to make sound decisions about the environment. However, because we have little evidence to date that students develop critical thinking skills through Citizen Science, it is not possible to say whether students who have participated in such programs become better decision-makers relative to the environment.

In contrast, the goals of environmental action programs such as Earth Force focus more explicitly on environmental and civic attitudes and citizenship skills, with the science investigation component being secondary (Earth Force 2003). In practice, the distinction between programs that focus on scientific literacy and those that focus on environmental action should be viewed more as a continuum than a dichotomy, as some Citizen Science programs also include components related to policy and local action. For example, students in CLO's Classroom FeederWatch have transformed schoolyards into wildlife habitats, and participants in The Birdhouse Network create nest sites and bluebird trails. Garden Mosaics may be viewed as a program that combines Citizen Science with a more explicit community action focus.

Future research might focus on the development of critical thinking and environmental decision-making in youth participants in Citizen Science programs. In addition, future studies might want to determine whether programs such as Garden Mosaics and Earth Force, which attempt to bridge science research and activism, are able to meet some of the behavioral as well as science education goals of environmental education.

ACKNOWLEDGMENTS

The Citizen Science Program at CLO has received generous funding from the Directorate for Education and Human Resources at the National Science Foundation: "Citizen Science Online" (ESI-0087760); "Parents Involved/Pigeons Everywhere" (ESI-9802248); "Schoolyard Ornithology Resource Project" (ESI-9618945); "Cornell Nest Box Network" (ESI-9627280); "Project BirdWatch" (ESI-9550541); "Public Participation in Ornithology" (ESI-9155700). We also acknowledge support from the National Fish and Wildlife Foundation, Florence and John Schumann Foundation, Wallace Genetic Foundation, Environmental Protection Agency, US Department of Agriculture Forest Service, and US Fish & Wildlife Service.

Garden Mosaics originally received funding from the US Department of Agriculture Northeast Sustainable Agriculture Research and Education Program. Garden Mosaics has received generous funding from the Directorate for Education and Human Resources at the National Science Foundation to expand the program nationally (ESI-0125582). We also acknowledge support from the College of Agriculture and Life Sciences at Cornell University, and we thank Rebekah Doyle, Gretchen Ferenz, Alan Berkowitz, Stephanie Thompson, and the other Garden Mosaics Leadership Team members.

REFERENCES

Barstow, D. and Diarra, C. (1997). Mars exploration: students and scientists working together. In *Internet Links for Science Education: Student–Scientist Partnerships*, ed. K. C. Cohen. New York, NY: Plenum Press, pp. 111–132.

Becker, M. L., Congalton, R. G., Budd, R, and Fried, A. (1998). A GLOBE collaboration to develop land cover data collection and analysis protocols. *Journal of Science Education and Technology*, **7**(1): 85–96.

Bonney, R. and Dhondt, A. A. (1997). Project FeederWatch. In *Internet Links to Science Education: Student Scientist Partnerships*, ed. K. C. Cohen. New York, NY: Plenum Press, pp. 31–54.

Brooks, D. R. (2000). *Spacecraft-based Earth Science in the 21st Century: Opportunities and Challenges for Student–Scientist Partnerships*. Workshop on Environmental Education at the School Level in the Field of Air Quality. Santori, Greece. [Online] URL: http://www.mcs.drexel.edu/~dbrooks/globe/santorini/santorini_paper.htm.

Brooks, D. R., Schanbacker, K., and Suchanic, D. (1999). *GLOBE and Integrated Pest Management: A case study for developing science partnerships*. Fourth Annual GLOBE Conference, Durham, NH. [Online] URL: http://www.globe.gov/fsl/html/templ_unh1999.cgi? Brooks-IPM&lang=en&nav=1.

Carlsen, W. S., Trautmann, N. M., Cunningham, C. M., and Krasny, M. E. (2004). *Watershed Dynamics*. Arlington, VA: National Science Teachers Association.

Chakane, M. and Trumbull, D. J. (2000). *SORP Evaluation: Classroom Implementation, Teacher Workshops, and Students' Inquiry Projects During the 1999/2000 Academic Year*. Department of Education, Cornell University.

Chambers, R. (1999). *Whose Reality Counts?: Putting the First Last*. London: Intermediate Technology Publications.

Cohen, K. C., ed. (1997). *Internet Links for Science Education: Student–Scientist Partnerships*. New York: Plenum Press.

Doyle, R. and Krasny, M. E. (2003). Participatory Rural Appraisal as an approach to environmental education in urban community gardens. *Environmental Education Research*, **9**(1): 91–115.

Earth Force. (2003). [Online] URL: http://www.earthforce.org/.

Freudenberger, K. S. (1999). *Rapid Rural Appraisal (RRA) and Participatory Rural Appraisal (PRA): a manual for CRS field workers and partners*. Baltimore, MD: Catholic Relief Services.

Gaventa, J. (1988). Participatory research in North America. *Convergence*, **21**(2/3), 19–27.

Gerber, J. M. (1992). Farmer participation in research: a model for adaptive research and education. *American Journal of Alternative Agriculture*, **7**(3), 118–121.

GLOBE. (2003). [Online] URL: http://www.globe.gov.

Hames, R. S., Rosenberg, K. V., Lowe, J. D., Barker, S. E., and Dhondt, A. A. (2002). Adverse effects of acid rain on the distribution of the Wood Thrush *Hylocichla mustelina* in North America. *Proceedings of the National Academy of Sciences of the United States of America*, **99**(17), 11235–11240.

Harnik, P. G. and Ross, R. M. (2003a). Developing effective K-12 geoscience research partnerships. *Journal of Geoscience Education*, **51**(1), 5–8.

(2003b). Assessing data accuracy when involving students in authentic paleontological research. *Journal of Geoscience Education*, **51**(1), 76–84.

Hartup, B. K., Bickal, J. M., Dhondt, A. A., Ley, D. H. and Kollias, G. V. (2001). Dynamics of conjunctivitis and *Mycoplasma gallisepticum* infections in House Finches. *Auk*, **118**, 327–333.

Hochachka, W. M. and Dhondt, A. A. (2000). Density-dependent decline of host abundance resulting from a new infectious disease. *Proceedings of the National Academy of Sciences*, **97**, 5503–5306.

Hochachka, W. M., Wells, J. V., Rosenberg, K. V., Tessaglia-Hymes, D. L., and Dhondt, A. A. (1999). Irruptive migration of Common Redpolls. *Condor*, **101**, 195–204.

Ison, R. L. and Russell, D. (1999). *Agricultural Extension and Rural Development: Breaking Out of Traditions*. Cambridge: Cambridge University Press.

Jordan, N., White, S., Gunsolus, J., Becker, R., and Damme, S. (2000). Learning groups developing collaborative learning methods for diversified, site-specific weed management: a case study from Minnesota, USA. In *Cow Up a Tree: Knowing and Learning Processes for Change in Agriculture. Case Studies from Industrial Countries*, ed. M. Cerf, D. Gibbon, B. Hubert, R. Ison, J. Jiggons, M. Paine, J. Proost, and N. Roling. Paris, France: INRA, pp. 85–95.

Krasny, M. E. (1999). Reflections on nine years of conducting high school research programs. *Journal of Natural Resources and Life Sciences Education*, **28**, 1–7.

Krasny, M. E. and Doyle, R. (2002). Participatory approaches to extension in a multi-generational, urban community gardening program. *Journal of Extension*, **40**(5). [Online] URL: http://www.joe.org/joe/2002october/a3.shtml.

Krasny, M. E. and Lee, S. K. (2002). Social learning as an approach to environmental education: lessons from a program focusing on non-indigenous, invasive species. *Environmental Education Research* **8**(2), 101–119.

Krasny, M. E., Berger, C., and Welman, A. (2001). *Long Term Ecological Research Educator's and Student's Manual*. Ithaca, NY: Department of Natural Resources. [Online] URL: http://www.dnr.cornell.edu/ext/LTER/lter.html.

Krasny, M. E., Trautmann, N. M., Carlsen, W. S., and Cunningham, C. M. (2002). *Invasion Ecology*. Arlington, VA: National Science Teachers Association.

LaBranche, M., Bonney, R. E., and Connelly, V. (2000). Distributions of Rock Dove color morphs described from data submitted by project volunteers. Joint Millennial Meeting of American Ornithologists' Union, British Ornithologists' Union, Society of Canadian Ornithologists, St. Johns, Newfoundland, Canada.

Lawless, J. G. and Rock, B. N. (1998). Student Scientist Partnerships and data quality. *Journal of Science Education and Technology*, **7**(1), 9–13.

Lederman, N. G. (1998). The state of science education: subject matter without context. *Electronic Journal of Science Education*, **3**(2). [Online] URL: http://unr.edu/homepage/jcannon/ejse/lederman.html.

Lewenstein, B. (2001). *PIPE Evaluation Report. Year 2, 1999–2000*. Ithaca, NY: Department of Communication, Cornell University.

Long-Term Ecological Research. (2003). [Online] URL: http://lternet.edu/.

Melchior, A. and Bailis, L. N. (2003). *2001–2002 Earth Force Evaluation: Program Implementation and Impacts*. Waltham, MA: Center for Youth and Communities, Heller Graduate School, Brandeis University.

Millar, J. and Curtis, A. (1999). Challenging the boundaries of local and scientific knowledge in Australia: opportunities for social learning in managing temperate upland pastures. *Agriculture and Human Values*, **16**, 389–399.

Mordock, K. and Krasny, M. E. (2001). Participatory Action Research: a theoretical and practical framework for environmental education. *Journal of Environmental Education*, **32**(3), 15–20.

Moss, D. M., Abrams, E. D., and Kull, J. A. (1998). Can we be scientists too? Secondary student perceptions of scientific research from a project-based classroom. *Journal of Science Education and Technology*, **7**(2), 149–161.

National Audubon Society. (2003). *Christmas Bird Count*. [Online] URL: http// www.Audubon.org/bird/cbc.

National Research Council (NRC). (1996). *National Science Education Standards*. Washington, DC: National Academy Press.

North American Association for Environmental Education (NAAEE). (1999). *Excellence in Environmental: Education Guidelines for Learning (K-12)*. [Online] URL: http://naaee.org/npeee/learner_guidelines.php (retrieved January 2004).

Pollock, K. H., Jones, C. M., and Brown, T. L. (1994). Angler surveys and their applications in fisheries management. *American Fisheries Society: Special Publication*, **25**, 371 pp.
Rock, B. N. and Lauten, G. N. (1996). K-12th grade students as active contributors to research investigations. *Journal of Science Education and Technology*, **5**(4), 255–266.
Roling, N. and Wagemakers, A. (1998). *Facilitating Sustainable Agriculture: Participatory Learning and Adaptive Management in Times of Environmental Uncertainty*. New York, NY: Cambridge University Press.
Rosenberg, K. V., Lowe, J. D., and Dhondt, A. A. (1999). Effects of forest fragmentation on breeding tanagers: a continental perspective. *Conservation Biology*, **13**, 568–583.
Ross, R. M., Harnik, P. G., Almon, W. D., Sherpa, J. M., Goldman, A. M., Nester, P. L., and Chiment, J. J. (2003). The mastodon matrix project: an experiment with large-scale public collaboration in paleontological research. *Journal of Geoscience Education*, **52**(1), 39–47.
Saldivar-Tanaka, L. and Krasny, M. E. (2005). The role of NYC Latino community gardens in community development, open space, and civic agriculture. In press, *Agriculture and Human Values*.
SRI International. (1997). *GLOBE Year 2 Evaluation Report: Implementation and Progress*. [Online] URL: http://www.globe.gov/fsl/evals/y2full.pdf
(1998). *GLOBE Year 3 Evaluation Report: Implementation and Progress*. [Online] URL: http://www.globe.gov/fsl/evals/y3full.pdf
(2000). *GLOBE Year 5 Evaluation Report: Classroom Practices*. [Online] URL: http://www.globe.gov/fsl/evals/y5full.pdf
(2002). *GLOBE Year 6 Evaluation Report: Explaining Variation in Implementation*. [Online] URL: http://www.globe.gov/fsl/evals/y6full.pdf
TERC. (1997). *National Conference on Student & Scientist Partnerships*. Cambridge, MA: TERC and Concord Consortium.
Thompson, S. (2002). *Garden Mosaics Evaluation Report*. Ithaca, NY: Seavoss, Inc.
Tinker, R. (1997). Student Scientist Partnerships: shrewd maneuvers. In *Internet Links for Science Education: Student–Scientist Partnerships*, ed. K. C. Cohen. New York: Plenum Press, pp. 5–16.
Trautmann, N. M., Carlsen, W. S., Krasny, M. E., and Cunningham, C. M. (2001). *Assessing Toxic Risk*. Arlington, VA: National Science Teachers Association.
Trautmann, N. M., Krasny, M. E., Carlsen, W. S., and Cunningham, C. M. (2003). *Decay and Renewal*. Arlington, VA: National Science Teachers Association.
Trumbull, D. J., Grudens-Schuck, N., and Scarano, G. (1998). *Evaluation of CFW Use in the 1997/98 Academic Year*. Ithaca, NY: Department of Education, Cornell University.
Trumbull, D. J., Scarano, G., and Chakane, M. (1999a). *SORP Formative Evaluation Report*. Ithaca, NY: Department of Education, Cornell University.
(1999b). *SORP Formative Evaluation Report #2*. Ithaca, NY: Department of Education, Cornell University.
Trumbull, D. J., Bonney, R., Bascom, D., and Cabral, A. (2000). Thinking scientifically during participation in a citizen-science project. *Science Education*, **84**, 265–275.
Wells, J. V., Rosenberg, K. V., Dunn, E. H., Tessaglia, D. L., and Dhondt, A. A. (1998). Feeder counts as indicators of spatial and temporal variation in winter abundance of resident birds. *Journal of Field Ornithology*, **69**, 577–586.
Wuest, S. B., McCool, D. K., Miller, B. C., and Veseth, R. J. (1999). Development of more effective conservation farming systems through participatory on-farm research. *American Journal of Alternative Agriculture*, **14**(3), 98–102.

Index